HANDBOOK OF ELECTRONIC MATERIALS
Volume 9

HANDBOOK OF ELECTRONIC MATERIALS

Compiled by:
ELECTRONIC PROPERTIES INFORMATION CENTER
Hughes Aircraft Company
Culver City, California

Sponsored by:

U.S. DEFENSE SUPPLY AGENCY
Defense Electronics Supply Center
Dayton, Ohio

HANDBOOK OF ELECTRONIC MATERIALS
Volume 9

Electronic Properties of Composite Materials

Maurice A. Leeds
Electronic Properties Information Center
Hughes Aircraft Company, Culver City, California

IFI/PLENUM · NEW YORK-WASHINGTON-LONDON · 1972

This document has been approved for public release and sale;
its distribution is unlimited. Sponsored by U.S. Defense Supply
Agency, Defense Electronics Supply Center, Dayton, Ohio.
Under Contract No. DSA 900-72-C-1182

Library of Congress Catalog Card Number 76-147312
ISBN 978-1-4615-9614-1   ISBN 978-1-4615-9612-7 (eBook)
DOI 10.1007/978-1-4615-9612-7
©1972   IFI/Plenum Data Corporation, a Subsidiary of
Softcover reprint of the hardcover 1st edition 1972
Plenum Publishing Corporation
227 West 17th Street, New York, N.Y. 10011

United Kingdom edition published by Plenum Press, London
A Division of Plenum Publishing Company, Ltd.
Davis House (4th Floor), 8 Scrubs Lane, Harlesden, NW10 6SE, London, England

CONTENTS

CONTENTS (CONT'D)

INTRODUCTION

Composites are the fastest growing class of structural material. Consequently, electronic properties are often difficult to find. This report was prepared in order to present a compilation of reliable data on the electronic and electrical properties of composites.

Composites provide an opportunity to tailor the properties to the application; a factor that allows designers an unlimited variety of new materials for new uses. It is this feature that has contributed to the rapid growth of composites.

The electrical properties of a composite can be of vital importance in the use or application of the material in a system. The designer therefore, must be able to obtain the necessary electrical or electronic property data to guide him in the materials selection. It is the purpose of this report to assist the designer and engineer in fulfilling that requirement.

Properties This report provides a compilation of the most commonly required electronic properties data of structural composites. Thermal properties often influence electrical design; consequently several of these properties are included. The specific properties tabulated are:

Arc Resistance	Thermal Conductivity
Arc Tracking Resistance	Linear Thermal Expansion
Dielectric Constant	Coefficient
Dissipation Factor	
Electrical (Volume) Resistivity	
Electrical (Volume) Conductivity	

Other electrical and thermal properties are compiled as the data was made available.

Materials The first requirement of a material for inclusion was compliance with the definition adopted for composites. Structural composites are defined as:

> A homogeneous combination of two or more materials, resulting in structural properties of the composite superior to those of either constituent. Each phase shall be identifiable on a microscopic or greater scale.

Further classification was accomplished by the list of general types included and specifically excluded as follows:

1

Included	Excluded
Fiber/Polymer Matrix	Metal Alloys
Fiber/Metal Matrix	Non-structural composites
Fiber/Ceramic Matrix	Solid State Electronic Devices
Lamellates	Electrical Contact Materials
Aligned Eutectics	
Dispersion Strengthened Alloys	
Mixtures (high impact strength)	

A few other classes were included though not listed above because no electrical data were available, i.e. Whisker/Matrix and Flake/Polymer Matrix.

When selecting a material for this report, a major consideration was that it be primarily useful as a structural component and that electrical properties be secondary. For example, concrete or asphalt for a highway is a structural material. However, the electronic properties become important when high frequency electronic detectors are embedded in order to monitor and control traffic. Boron and graphite fiber/polymer matrix composites are excellent structural materials, used in many applications including aircraft. In this regard, electrical properties become very important with the possibility of lightning strikes. Tungsten wire reinforced copper was developed as a high strength material suitable for structural applications (Reference 47). However, the excellent conductivity of this composite makes it a suitable contender for power transmission.

Occasionally, composite materials with a borderline compliance to the above criteria are included. This resulted when unusual or unique materials were uncovered during the search and the electronic properties were available. On the other hand, in a class of composites having a large population of similar materials, a few representative examples were selected from more than one readily available source.

In addition to the data itself, other information is provided. With most citations a brief statement clarifying important parameters is given. Additional material descriptions seemed superfluous because it was assumed the engineer is familiar with the material he is considering. Standard test methods stated in the reference are included in the compilation because the author believes the value of the data is greatly enhanced when the test method is given. Non-standard test methods described in the reference are noted in tables by an "R" or foot note. Where a "U" appears in the table, or no reference to test methods is given, the reader may assume that they are not known.

Data Sources A variety of literature provided the data for this compilation. Included were reports from conferences, journals and periodicals, books, government research reports and sales literature from materials manufacturers.

ASSESSMENT OF DATA

The precision and accuracy* of the data in this report must be carefully weighed by designers and engineers when determining the degree of applicability to specific projects. Several factors that should be considered are discussed below.

Composites are multiphase materials with two or more constituents. Therefore, within a given class with more than one available source for the constituents, the composite properties depend upon the constituents used. In addition, it is highly probable that the composite's properties will vary from lot to lot of the constituents.

The process by which a composite is manufactured has many factors, such as time, temperature, pressure, etc., each of which could influence the properties. A feature of many composites is the ability to fabricate the final shape with few or no intermediate operations. The tooling used for this shaping could influence the properties. Very often property data is determined on specimens of standard shapes formed in special tools. The same material formed into a different shape may have different values for the same property. This phenomenon is characteristic of many composites because of the high incidence of anisotropy.

Measurement accuracy is dependent upon the test method, especially when determining electrical properties. Accuracy is also influenced by the care in application of the method and the equipment used.

Property data is also affected by other variables such as
- Thermal and environmental history
- Properties of constituents
- Distribution of constituents
- Presence of foreign matter including voids

Finally, reported values are influenced by material sampling techniques and statistical analysis methods.

This report should be useful to the designer and engineer as a guide to the electrical properties of composites. For design data, however, closer contact to the references and other sources is advisable.

Many new composites are finding expanded use, yet electrical property data is scarce. The references in this report identify the sources of recent and current activity for a rapidly expanding and changing information field.

* Precision and accuracy, as used here, are defined in ASTM E177.

CERAMIC MATRIX

ALUMINUM PHOSPHATE REINFORCED WITH 5-994 FIBERGLASS   Ref. 7

Fiberglass coated with Dow Corning
DL-805 silicone resin

Thickness	Dielectric Constant	Loss Tangent
030 inch	3.44	0.0062

United Aircraft Tungsten Core Filament (.0005" Diameter)

Mean value of resistance (9 measurements) 44.6 ohms.
Length of sample between edges of mercury pool 0.5259".
Diameter as measured (3 measurements) 0.00393".
The effective resistivity (assuming homogeneous filament)
2600µΩcm.

Effect of Current on Above Resistivity

Current in Fiber (m. amp)	Resistance of Fiber (ohms)
1.73	44.2
2.60	42.7
6.00	47.5
10.00	44.4

GRAPHITE AND CARBON   Ref. 54

	HMG-50 YARN	HMS TOW
DENSITY gm/cm^3	1.68	1.84
RESISTIVITY ohm-cm	1.35×10^{-3}	0.83×10^{-3}
SPECIFIC HEAT cal/gm/°C	0.3	0.3

	Unit	KCF-100	KGF-200
Carbon content	%	99.5	99.8
Electric resistance	ohm-cm	7.5×10^{-3}	3.0×10^{-3}
Thermal conductivity*	Kcal/m hr°C	15.4	30-40
Coefficient of thermal expansion	10^{-6}/°C	1.7	1.5

* Calculated by Lorentz Number.
  By Lorentz, thermal conductivity is a function of the absolute
  temperature and the electrical conductivity.

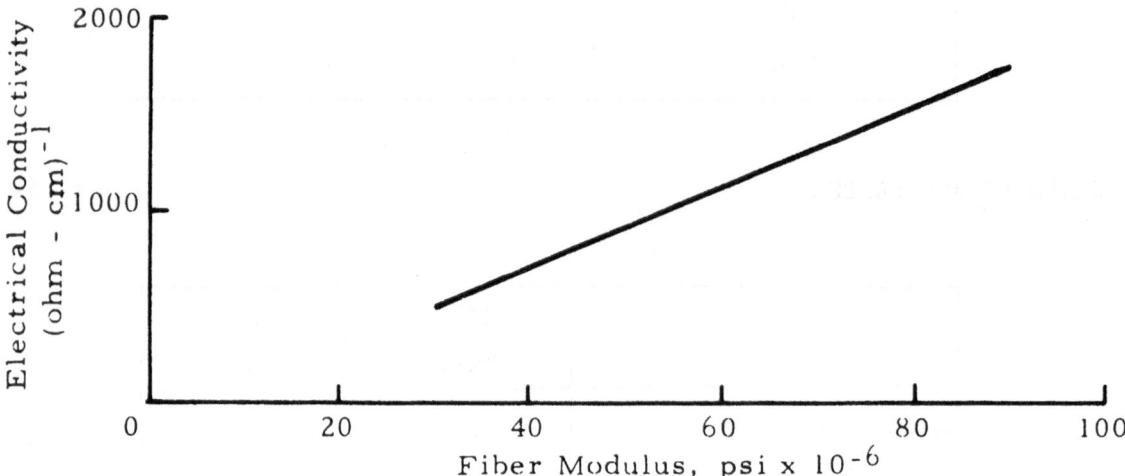

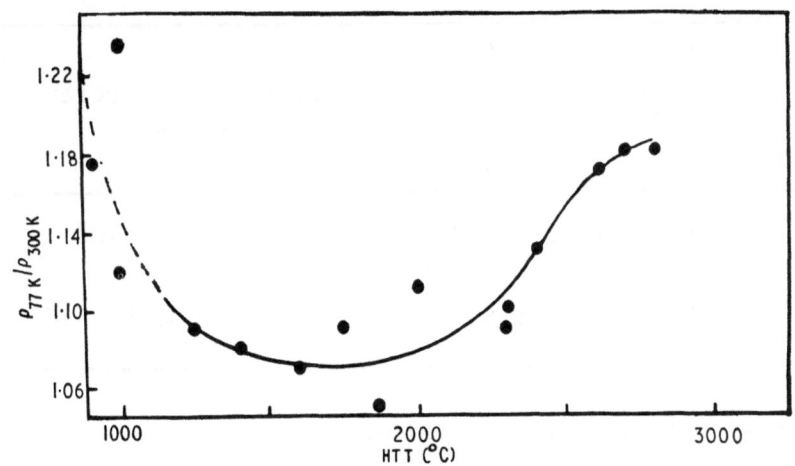

Variation of the resistivity ratio with HTT of specimen.

HTT - Heat treat temperature during graphitization.

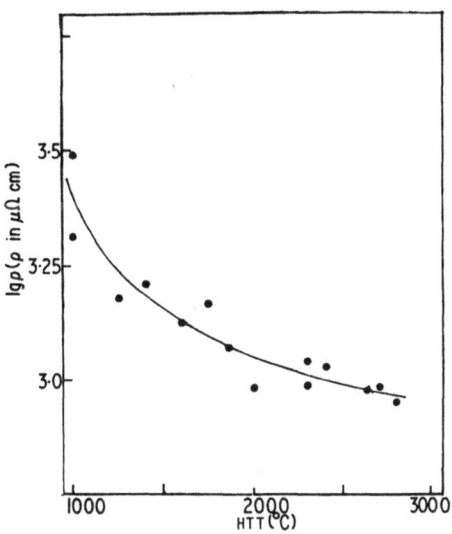

Variation of longitudinal resistivity for carbon fibers.

Material	Dielectric constant (Hz)	Dissipation factor tan δ (Hz)	Surface resistivity (ohm/cm)
"E" glass fibers	6.43 at 10^2 6.11 at 10^{10}	0.0042 at 10^2 0.006 at 10^{10}	10^{15}*
"A" glass fibers	6.8 at 5×10^5	0.007 at 5×10^5	10^{12}

* Estimate (figure for "E" glass not known).

GLASS FABRIC   Ref. 55

Frequency 1000 Hz

T°C	Dielectric Constant	Dissipation Factor
25	6.3	0.0037

Composition is 99.3% silica

Refrasil, Batt, Felt and Cloth

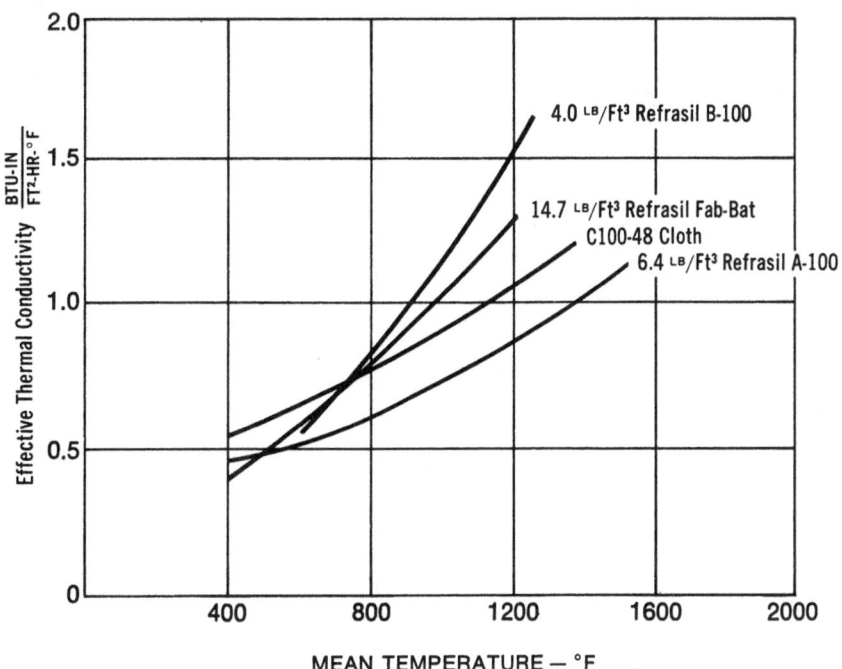

Sleeving and Cloth Breakdown Voltage

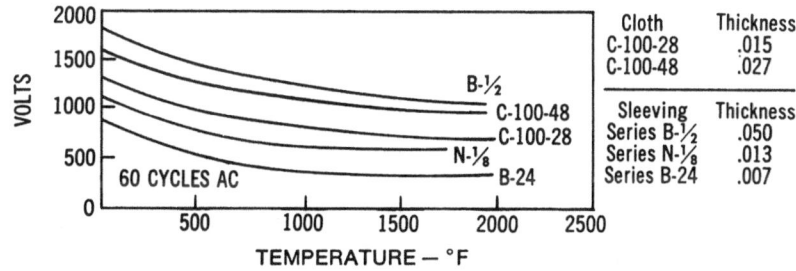

All Forms

Temp.	Volume Resistivity	Dielectric Constant	Power Factor	Frequency
80°F.	5 X 10^8		.001 — .004	60 cycle — 1 MC
	1 X 10^{14}*	1.3 — 2.3*		
500°F.	5 X 10^{12}	1.5	.002	60 cycle — 1 MC
900°F.	5 X 10^9	1.5	.001	1 KMC

*Depending on moisture content. The dielectric constant of pure non-porous silica is about 3.8.

Freq., 8.52 GHz

T°C	κ'	tan δ
25*	3.02	.0054
25	2.98	;0019
98	2.97	.0018
198	2.96	.0016
307	2.95	.0015
418	2.95	.0014
497	2.945	.0014
591	2.95	.0016
729	2.96	.0022
828	2.975	.0029
905	2.99	.0035
995	3.01	.0042

* As received, other values after vacuum bake
  for 24 hours at 125°C.

QUARTZ FABRIC   Ref. 55

Frequency 1000 Hz

T°C	Dielectric Constant	Dissipation Factor
25	3.7	0.0001

The composites consist of powdered iron calcium borate
glass and aluminum compacted, sintered and forged.

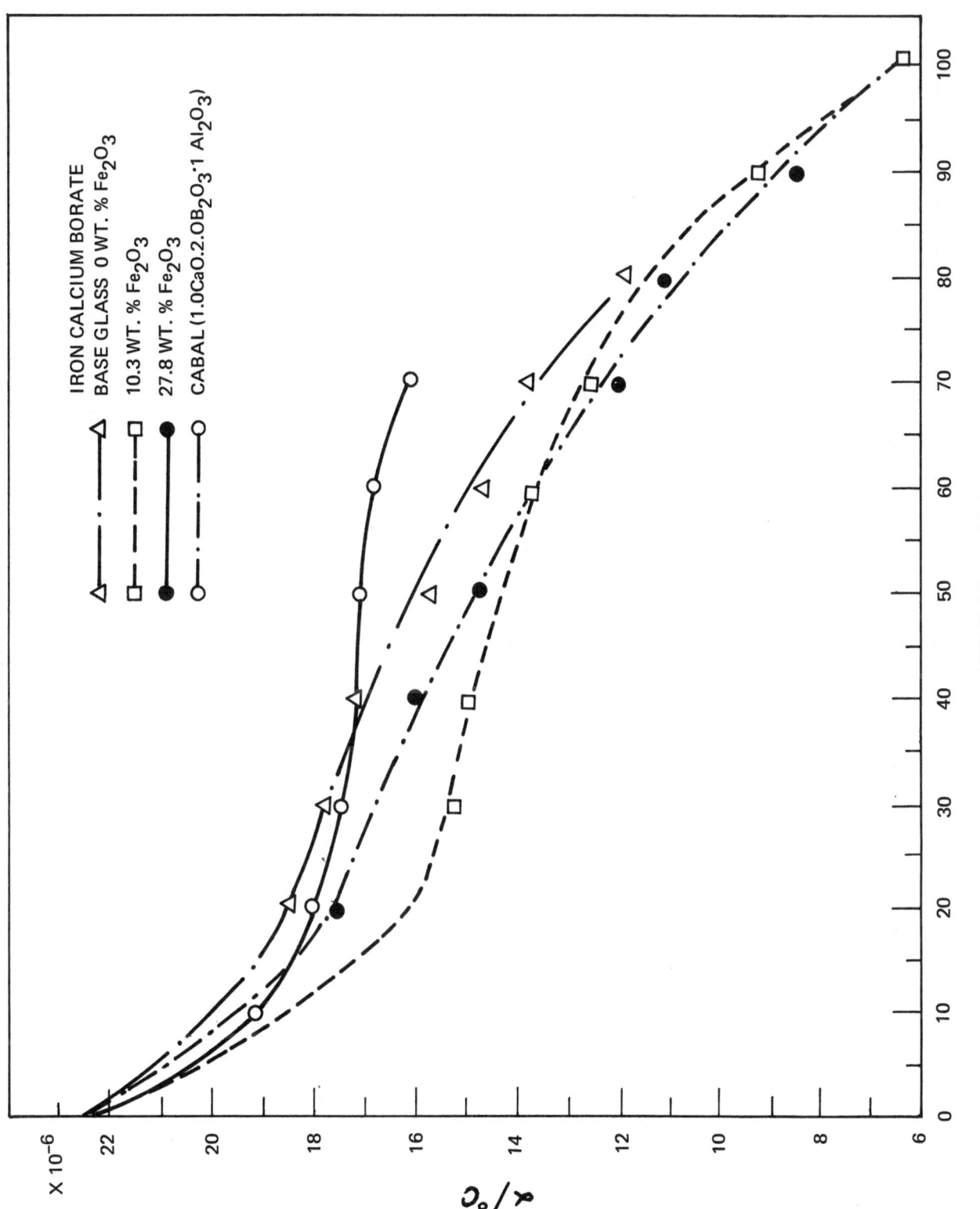

EFFECT OF GLASS CONTENT ON THE COEFFICIENT OF LINEAR EXPANSION OF COMPOSITES

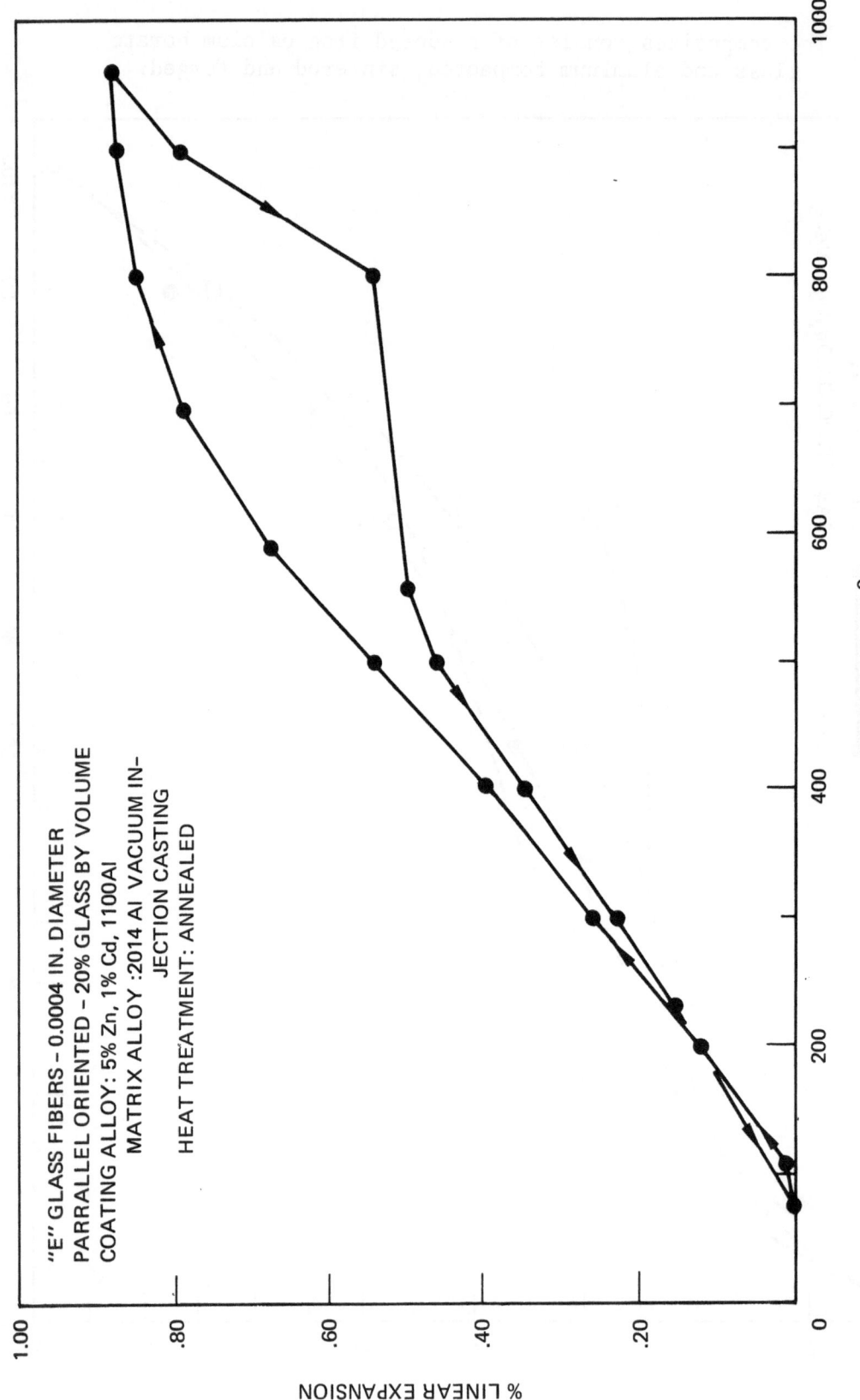

"E" GLASS FIBERS – 0.0004 IN. DIAMETER
PARRALEL ORIENTED – 20% GLASS BY VOLUME
COATING ALLOY: 5% Zn, 1% Cd, 1100Al
MATRIX ALLOY :2014 Al VACUUM IN-
JECTION CASTING
HEAT TREATMENT: ANNEALED

% LINEAR EXPANSION

TEMPERATURE,°F

Thermal Expansion Curve

Note: The unusual contraction of this composite is discussed in the reference.

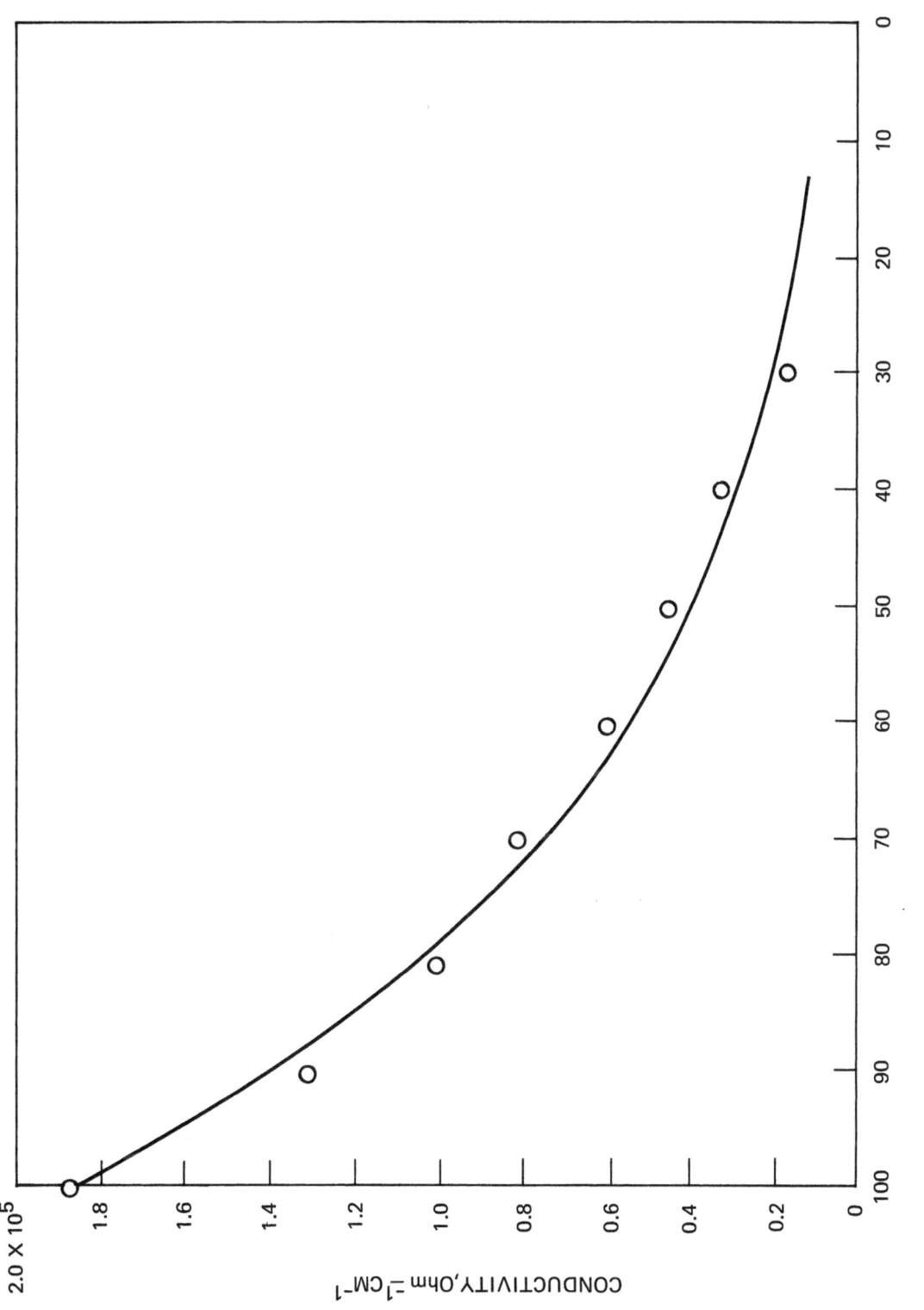

ALUMINIUM VOL. PERCENT

VARIATION OF ELECTRICAL CONDUCTIVITY
OF COMPOSITES WITH GLASS CONTENT

GLASS-METAL   Ref. 63

Glassed Steel

	Crystallized Glass Layer	Amorphous Glass Layer
Thermal conductivity		
Ceramic component only, Btu-in/hr ft^2°F	8	6
Composite, 0.05 in. ceramic plus 11/16 in. steel, k/1	123	98
Typical chemical service, (liquid product, steam jacketed vessel), overall "U"	87	74
Dielectric strength		
Volts per mil, room temp., 60 cycle	735	700
Dielectric constant		
f = 10^6 cps	8.24	8.02

Unidirectional Fiber Orientation

	Graphite-Epoxy [*]
Electrical resistivity (microhm-in.)	
Longitudinal	1.18×10^3
Transverse	2.76×10^6
Coefficient of thermal expansion (10^{-6} in./in./°F)	
Longitudinal	-0.2
Transverse	16
Thermal conductivity (Btu./hr./ft.^{2}/°F/ft.)	
Longitudinal	28
Transverse	0.8
Specific heat (Btu./lb./°F)	0.21

[*] Tested at room temperature.

Unidirectional Fiber Orientation

Sample Composites		K (W.m^{-1}.°C^{-1})	σ (ohm^{-1}.m^{-1}x10^2)	E (10^{10}N/m^2)	$\frac{K_L}{K_T}$	$\frac{\sigma_L}{\sigma_T}$	α@25°C (°C^{-1}x10^{-6})	α@100°C (°C^{-1}x10^{-6})	Density (kg.m^{-3}x10^{-3})	Voids %
40% Type I S. Treated	L	39.3	457	9.87	50	119	17.0	−1.06	1.38	5.8
	T	0.79	2.3	0.50				73.5		
50% Type I S. Treated	L	51.1	526		53	142	−	−1.18	1.48	3.4
	T	0.96	3.7				17.0	54.0		
60% Type I S. Treated	L	64.9	667	15.40	47	131	−	−1.22	1.57	1.9
	T	1.38	5.1	0.65			15.5	49.5		
50% Type I Untreated	L	66.1	556	11.70	75	214	−	−0.47	1.47	4.2
	T	0.88	2.6	0.30			19.5	57.0		
40% Type II S. Treated	L	9.6	253	6.70	16	211	−	−0.035	1.30	9.2
	T	0.58	1.2	0.46			17.0	73.5		
50% Type II S. Treated	L	11.30	347		17	112	−	−0.20	1.40	6.0
	T	0.67	3.1				17.0	54.0		
60% Type II S. Treated	L	12.6	413	9.80	18	71	−	−0.37	1.44	7.0
	T	0.71	5.8	0.60			15.5	49.5		
50% Type II Untreated	L	11.7	345	8.97	18	128	−	−0.18	1.41	5.1
	T	0.67	2.7	0.63			19.5	57.0		

Type I fiber, $E = 46.4 \times 10^6$ psi (31.9×10^{10} N/m^2)
Type II fiber, $E = 34.0 \times 10^6$ psi (23.5×10^{10} N/m^2)
Test methods in reference. Extensive graphic data in reference.

Unidirectional Fiber Orientation

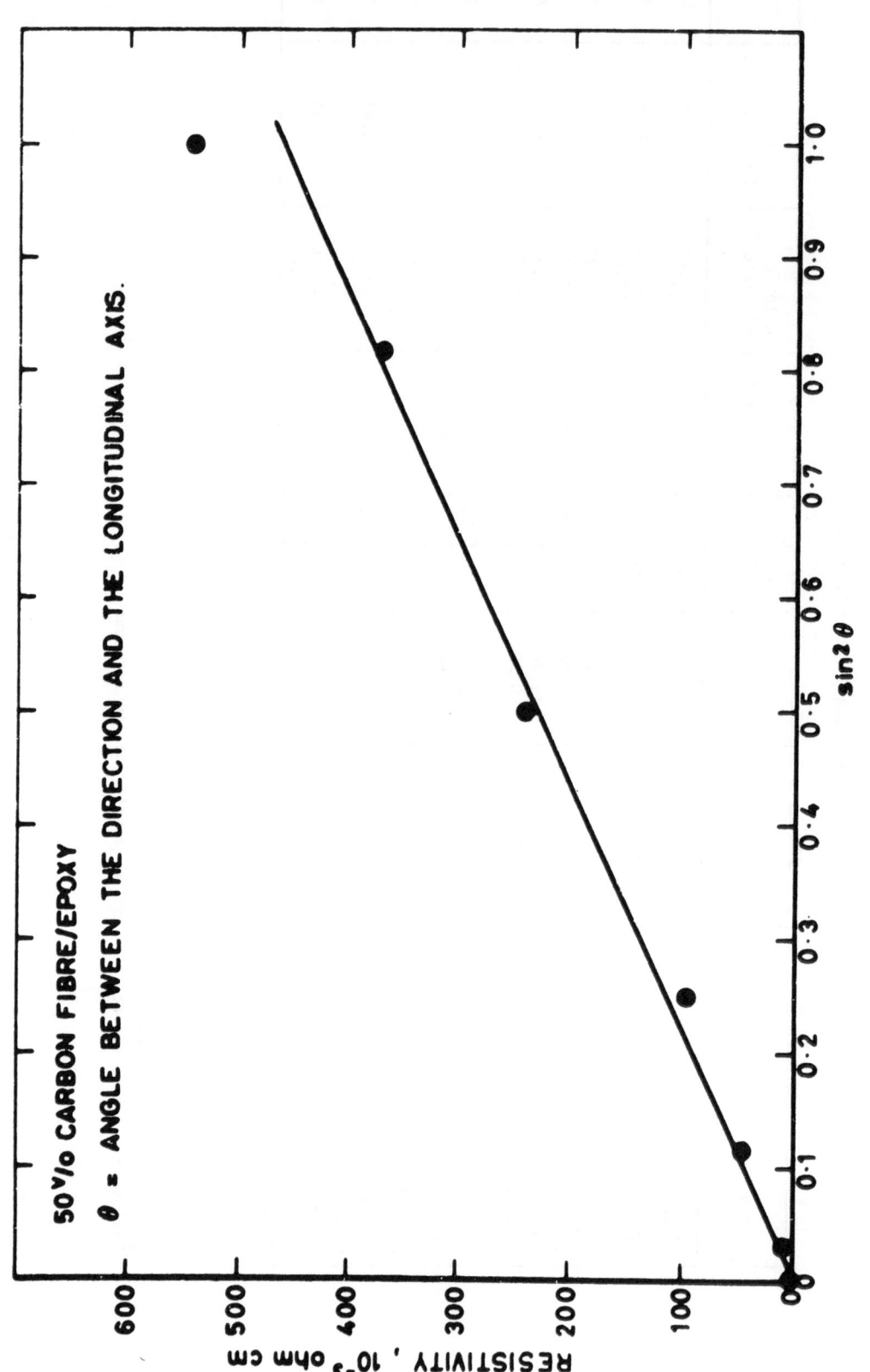

50°/o CARBON FIBRE/EPOXY

θ = ANGLE BETWEEN THE DIRECTION AND THE LONGITUDINAL AXIS.

PLOT OF ELECTRICAL RESISTIVITY VERSUS $\sin^2\theta$ FOR TYPE I COMPOSITE.

Test methods in reference.

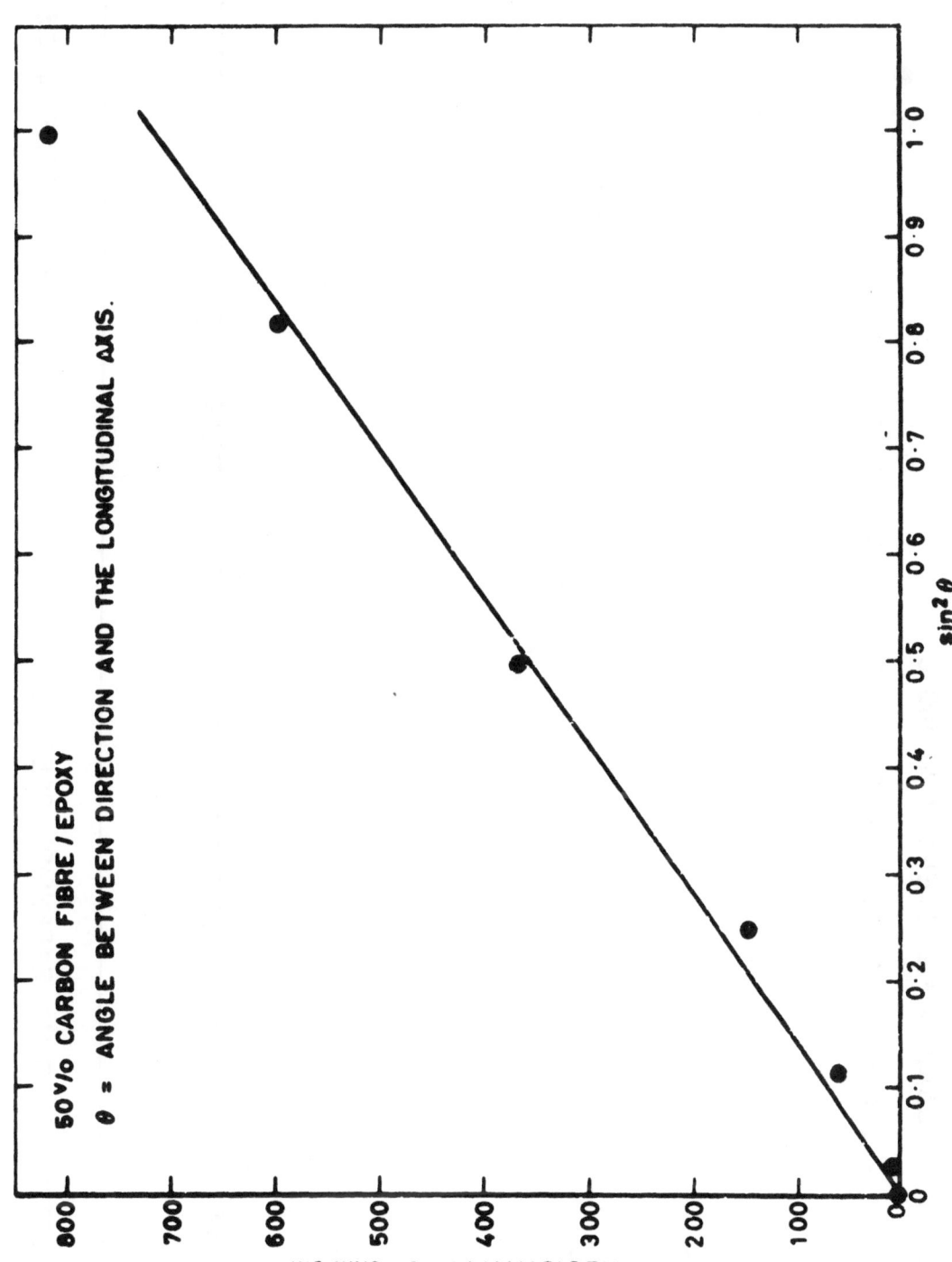

PLOT OF ELECTRICAL RESISTIVITY VERSUS sin²θ FOR TYPE II COMPOSITES.

Test methods in reference.

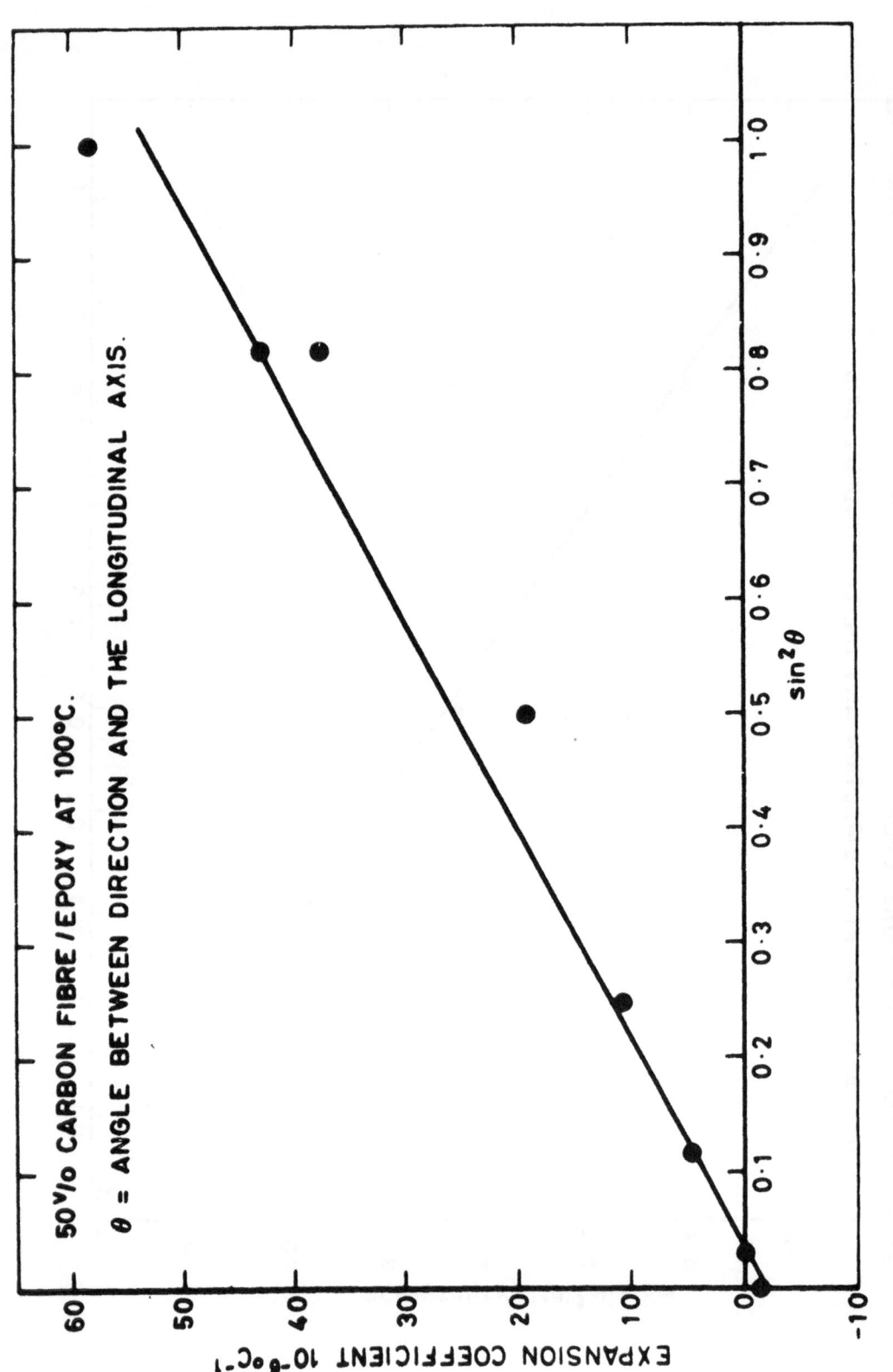

PLOT OF THERMAL EXPANSION COEFFICIENT VERSUS $\sin^2\theta$ FOR TYPE II COMPOSITES.

Test methods in reference.

CARBON/GRAPHITE-POLYESTER   Ref. 19

Unidirectional Fiber Orientation

	Modmor Type I Composite
Resistivity ohm-cm 10^{-6}	775 (25°C) 660 (180°C)
Thermal Conductivity	.04 (25°C)

BORON-EPOXY   Ref. 43

Unidirectional Fiber Orientation

	Boron-Epoxy *
Electrical resistivity (microhm-in.)	
Longitudinal	3.94×10^3
Transverse	3.94×10^{12}
Coefficient of thermal expansion (10^{-6} in./in./°F)	
Longitudinal	2.5
Transverse	7.9
Thermal conductivity (Btu./hr./ft.^{2}/°F/ft.)	
Longitudinal	1.1
Transverse	0.6
Specific heat (Btu.lb./°F	0.28

* Tested at room temperature.
  Test methods unknown.

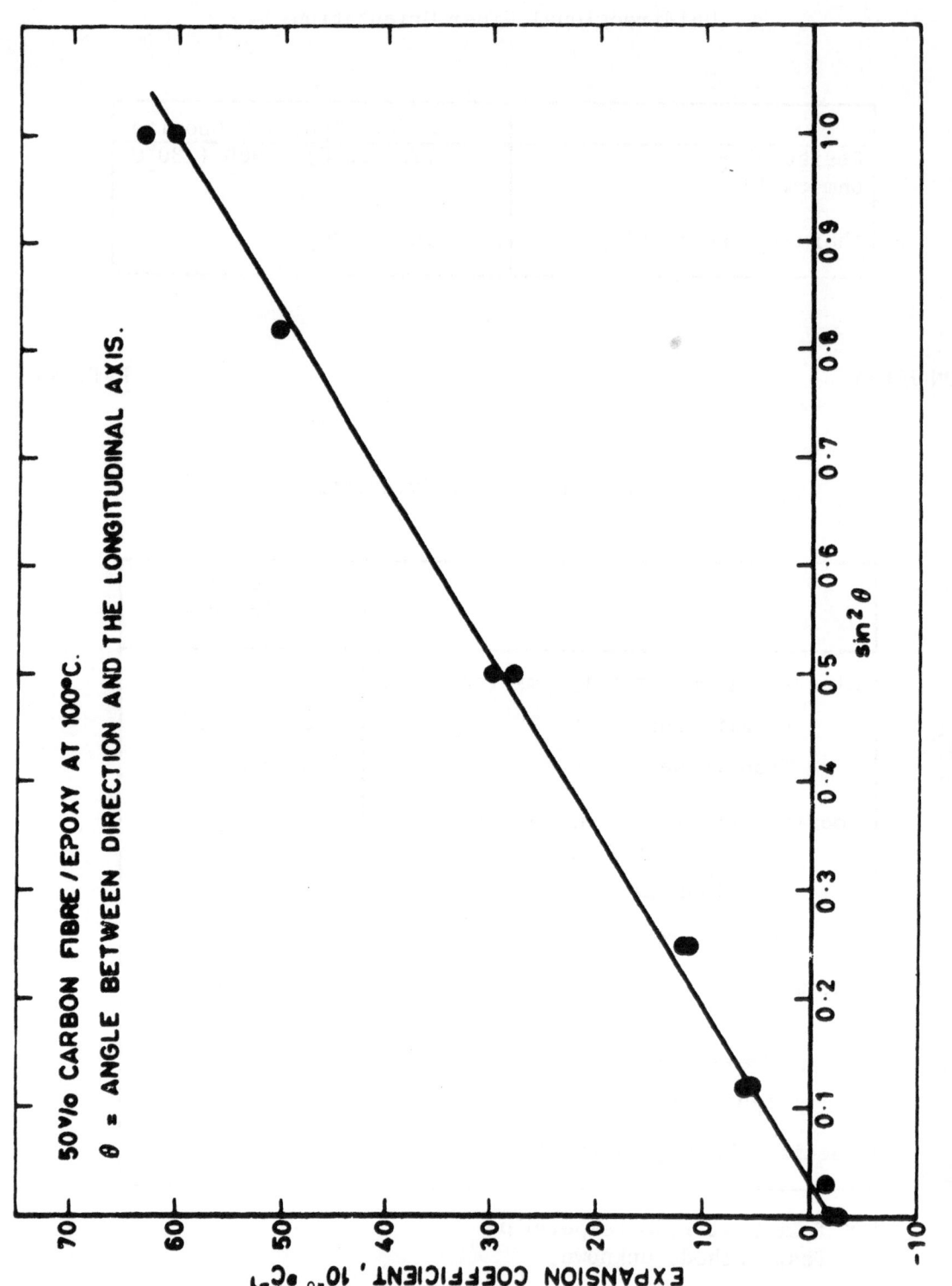

PLOT OF THERMAL EXPANSION COEFFICIENT VERSUS sin²θ FOR TYPE I COMPOSITES.

Test methods in reference.

Resin	Test Temperature, °F	Dielectric Constant, κ'	Dissipation Factor, tan δ	Loss Index
Amide-imide*	75	4.29	0.018	0.0772
	300	4.37	0.016	0.0699
	400	4.31	0.016	0.0690
	500	4.53	0.038	0.1721

FIBERGLASS-DIPHENYL OXIDE   Ref. 55

Resin	Test Temperature, °F	Dielectric Constant, κ'	Dissipation Factor tan δ	Loss Index
Diphenyl-oxide*	75	4.83	0.024	0.1159
	300	4.98	0.023	0.1145
	400	5.06	0.027	0.1366
	500	5.20	0.027	0.1404

FIBERGLASS-ACRYLIC   Ref. 55

Resin	Test Temperature, °F	Dielectric Constant, κ'	Dissipation Factor tan δ	Loss Index
Acrylic*	75	3.88	0.027	0.1048
	300	4.63	0.013	0.0602
	400	4.63	0.047	0.2176

* E glass style 181 fabric. Resin content approximately 35 percent.
  Test frequency 9.35 GHz.

Resin	Test Temperature °F	Dielectric Constant, κ'	Dissipation Factor, tan δ	Loss Factor
DAIP	75	4.16	0.010	0.0416
	300	4.21	0.014	0.0589
	400	4.24	0.017	0.0721
	500	4.29	0.029	0.1244

E glass style 181 fabric.   Resin content approximately 35 percent.
Test frequency 9.35 GHz.

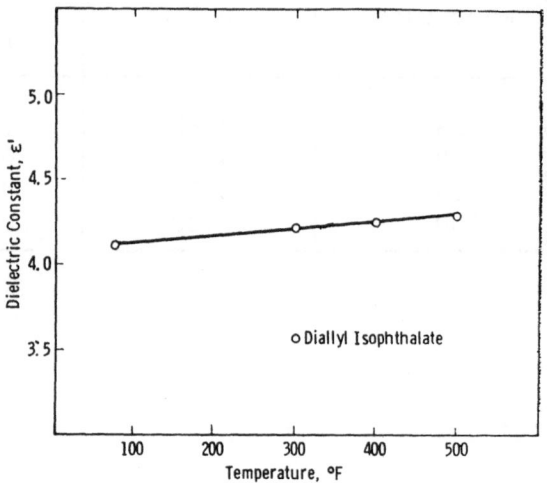

Dielectric constant at 9.35 GHz
of style-120 glass-fabric
laminates as function of
temperature (40 percent resin
content).

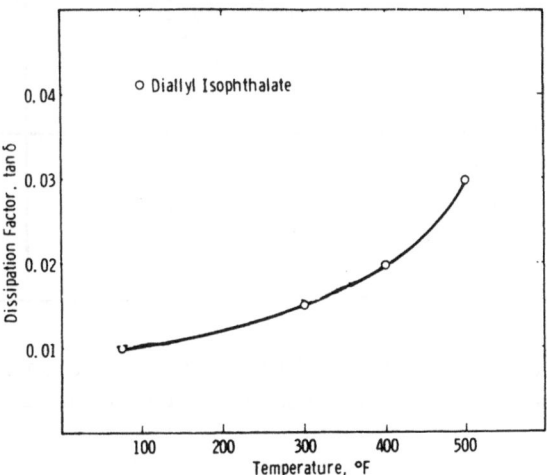

Dissipation factor at 9.35 GHz of
style-120 glass-fabric laminates as
function of temperature (40 percent
resin content).

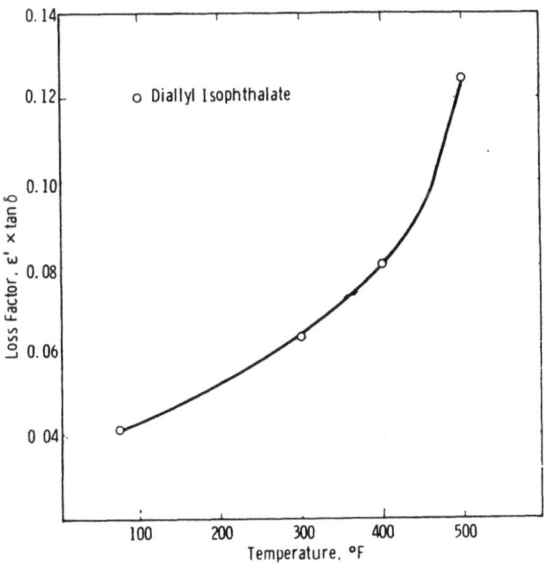

Loss factor at 9.35 GHz of style-120
glass-fabric laminates as function
of temperature (40 percent resin
content).

	Poly-Preg P 680*	Poly-Preg P 681**
Dielectric Constant		
9.3 GHz	3.9	
10^6 Hz		4.1
10^3 Hz		4.4
Dissipation Tangent		
9.3 GHz	0.016	0.014
10^6 Hz		0.013
10^3 Hz		

* Fiberglass style 181 fabric with I 550 finish
** Fiberglass style 1581 fabric with volan finish

FIBERGLASS-DIOLEFIN   Ref. 55

Resin	Test Temperature °F	Dielectric Constant, κ'	Dissipation Factor, tan δ	Loss Factor
Diolefin	75	3.41	0.0120	0.0409
	300	4.08	0.0104	0.0424
	400	4.20	0.0086	0.0361
	500	4.63	0.0068	0.0315

E glass style 181 fabric. Resin content approximately 35 percent.
Test frequency 9.35 GHz.

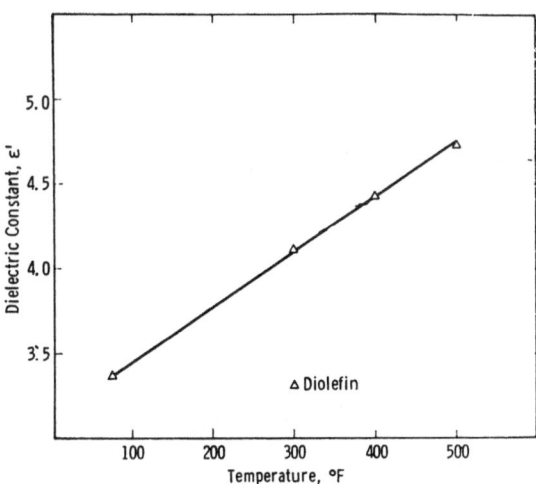

Dielectric constant at 9.35 GHz
of style-120 glass-fabric laminates
as function of temperature (40
percent resin content).

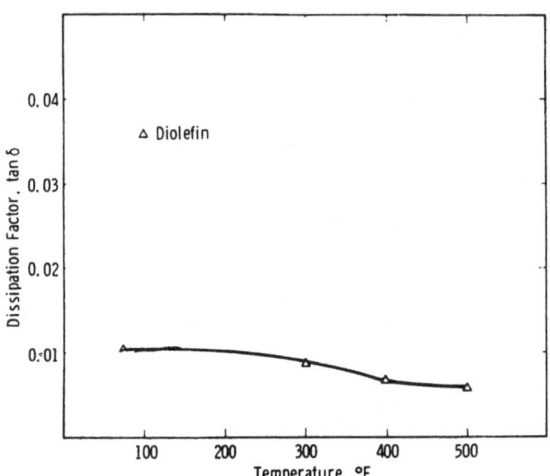

Dissipation factor at 9.35 GHz of
style-120 glass-fabric laminates as
function of temperature (40 percent
resin content).

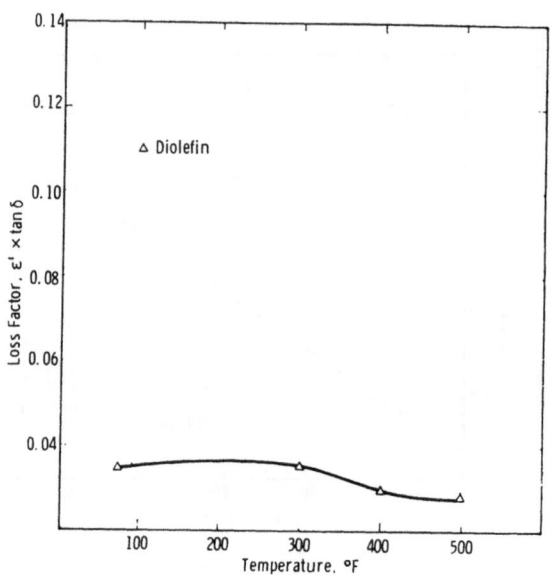

Loss factor at 9.35 GHz of style-120 glass-fabric laminates as function of temperature (40 percent resin content).

FIBERGLASS-NYLON

Effect of Silanes on Electrical Properties of
Glass** Reinforced Resins at 1000 CPS

Finish* on Fiberglass	Dielectric Constant		Dissipation Factor	
	Dry	16 Hrs 100% RH/120°F	Dry	16 Hrs 100% RH/120°C
None	4.5	8.3	0.035	5.2
Silane Y-5108	4.2	4.2	0.035	0.53
Silane Y-4087	4.7	5.2	0.037	0.50

  * All silanes used at 0.5 wt percent loading, based on glass
 ** 181 glass fabric reinforced resin, Glass content 40-45 % wet, %.

The fiberglass used is style 181 fabric.

Property	Test Method	F155 (28)	F161 (28)	Poly-Preg E-705 (60)	Poly-Preg E-741D (60)	Poly-Preg E-777 (60)	Poly-Preg E-780* (60)	F-153 (28)
Dielectric Constant 9.37 GHz Dry	MIL-R-9300	4.55	4.4		4.68	4.7		4.6
Dielectric Constant 9.37 GHz Wet	MIL-R-9300		4.4					
Dielectric Constant 10^6 Hz	U			4.76			4.86	
Dielectric Constant 10^3 Hz	U			4.82			5.15	
Dissipation Factor 9.37 Hz Dry	MIL-R-9300	0.010	.011		0.018	0.022		0.167
Dissipation Factor 9.37 Hz Wet	MIL-R-9300		.012					
Dissipation Factor 10^6 Hz	U			0.016			0.012	
Dissipation Factor 10^3 Hz	U			0.007			0.005	

* Flame resistant

33

Resin	Test Temperature, °F	Dielectric Constant, κ'	Dissipation Factor tan δ	Loss Index
Epoxy 1	75	4.06	0.015	0.0609
	300	4.30	0.033	0.1419
	400	4.39	0.037	0.1624
	500	4.44	0.042	0.1865
Epoxy 2	75	4.78	0.015	0.0717
	300	4.61	0.018	0.0830
	400	5.28	0.021	0.1109
	500	5.43	0.025	0.1358
Epoxy novolac 1	75	4.65	0.019	0.0884
	300	4.61	0.023	0.1060
	400	4.99	0.050	0.2495
	500			
Epoxy novolac 2	75	4.68	0.013	0.0608
	300	4.97	0.022	0.1093
	400	5.31	0.061	0.3239
	500			
Epoxy acrylate	75	4.59	0.025	0.1148
	300	5.16	0.042	0.2167
	400			
	500			

E glass style, 181 fabric. Resin content approximately 35 percent.
Test frequency 9.35 GHz.

S-Glass Fabric

Electrical resistivity (microhm-in.)	
Longitudinal	3.94×10^{19}
Transverse	3.94×10^{19}
Coefficient of thermal expansion (10^{-6} in./in./°F)	
Longitudinal	3.5
Transverse	11.4
Thermal conductivity (Btu-in./hr ft^2°F	
Longitudinal	0.17
Transverse	0.12
Specific heat (Btu./lb./°F)	0.24

Data collected at room temperature.

Ref. 61

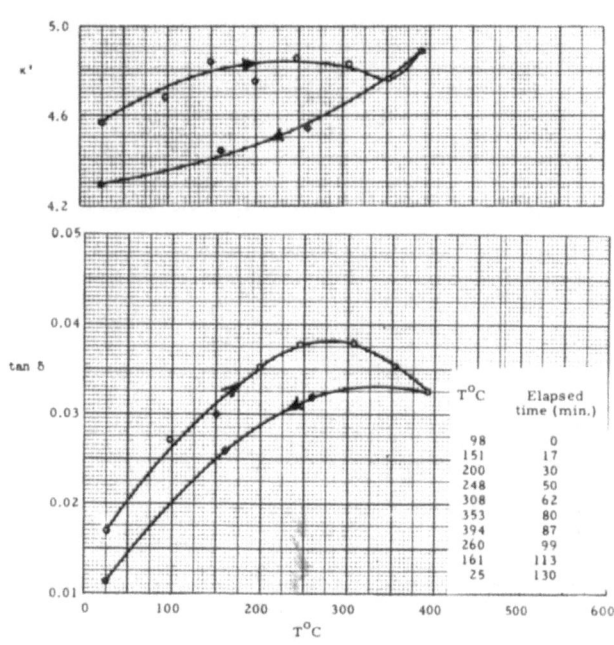

Fiberglass laminate with 181 glass cloth
tested at 8.52 GHz.

S-Glass Style 181 Fabric HTS Finish

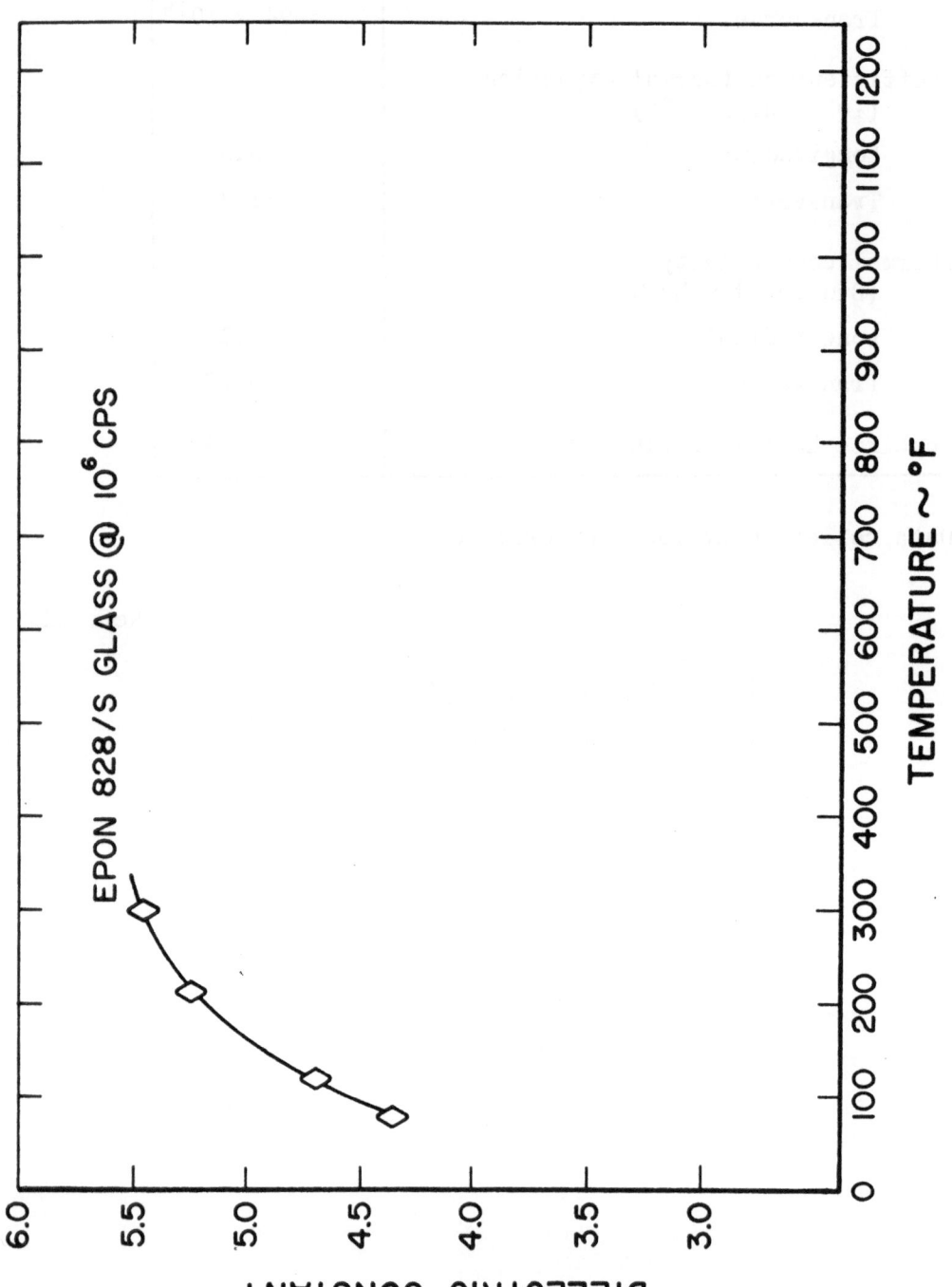

Effect of Temperature on Dielectric Constant of Epon 828 Composite

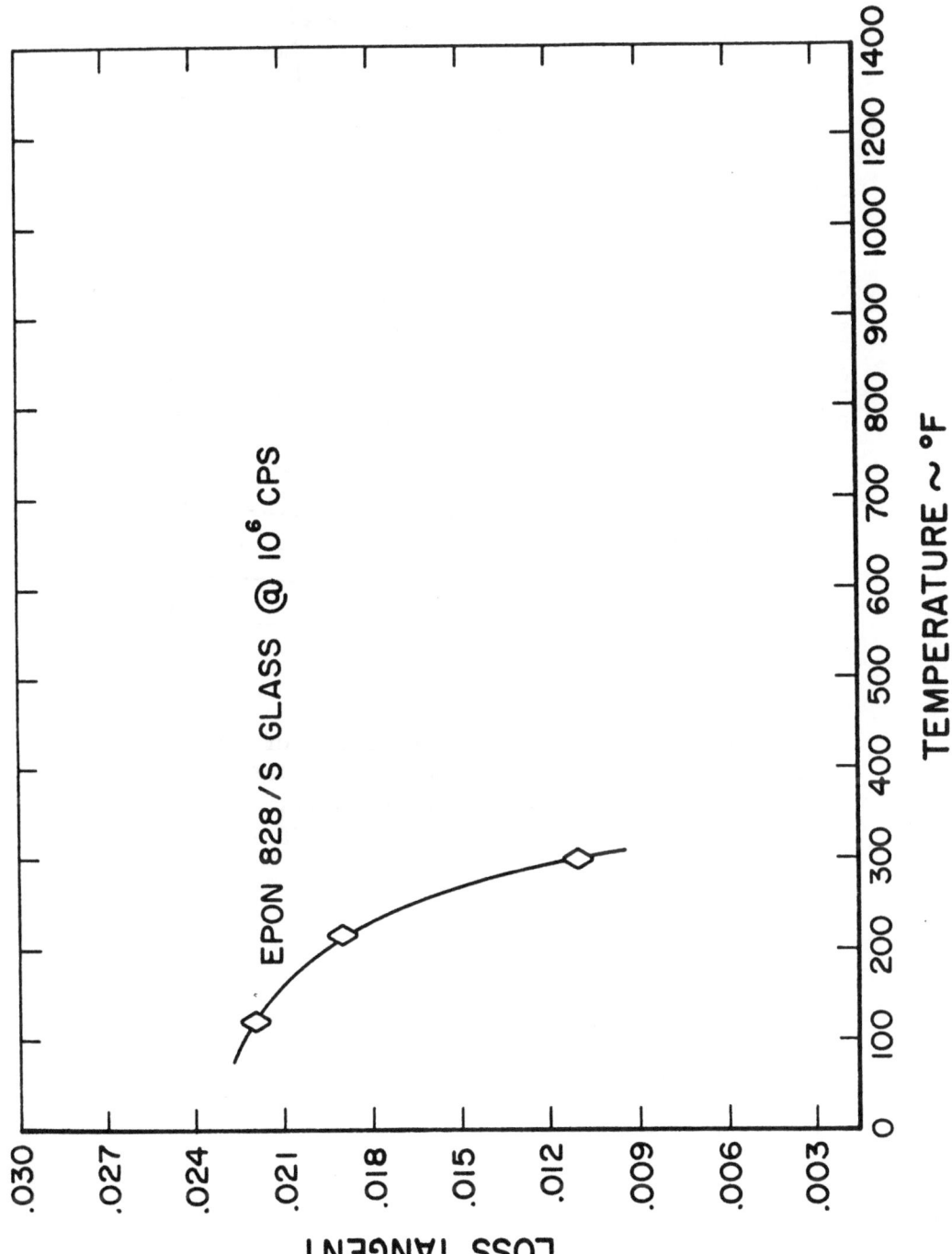

Effect of Temperature on Loss Tangent of EPON 828 Composite

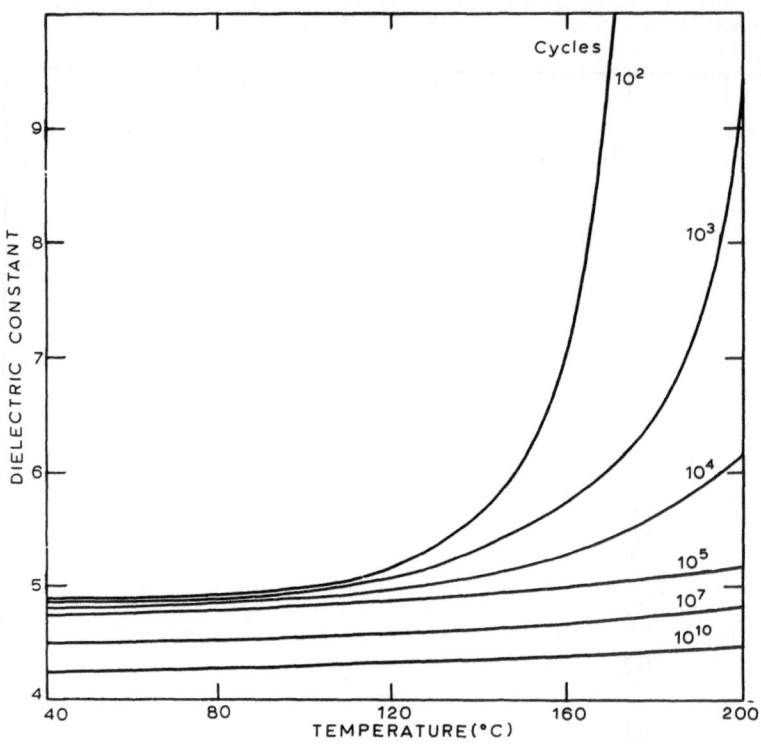

Dielectric constant vs. temperature of composite at various frequencies. Glass, "E" glass, 181 cloth; resin, epoxide, Epikote 828/BF_3-400, 2 p.h.r.; keying agent, Volan A.

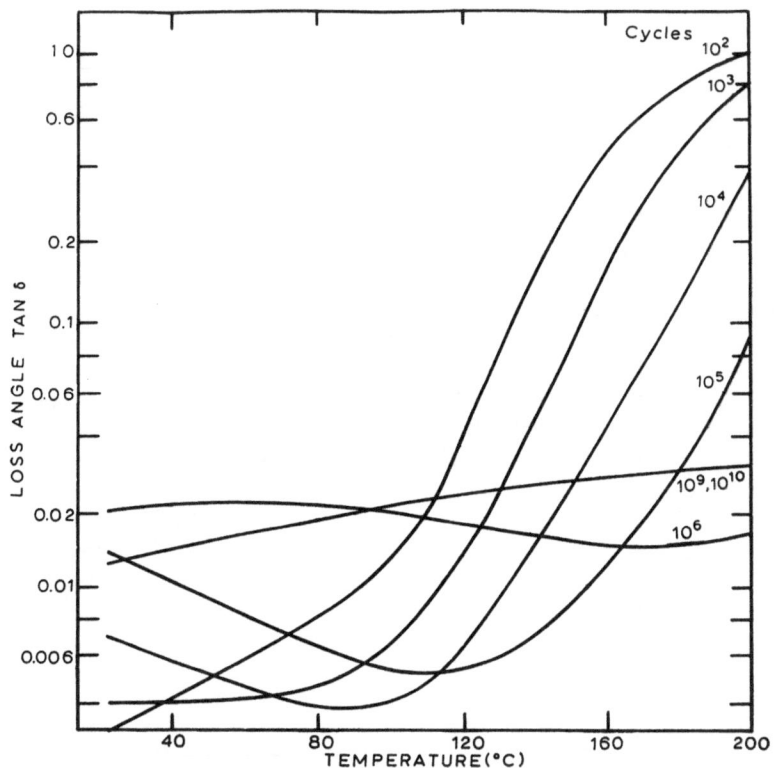

Loss angle, tan δ vs. temperature of
composites at various frequencies. Glass,
"E" glass cloth; resin, epoxide,
Epikote 828/BF$_3$-400, 3 p.h.r.; keying
agent, Volan A.

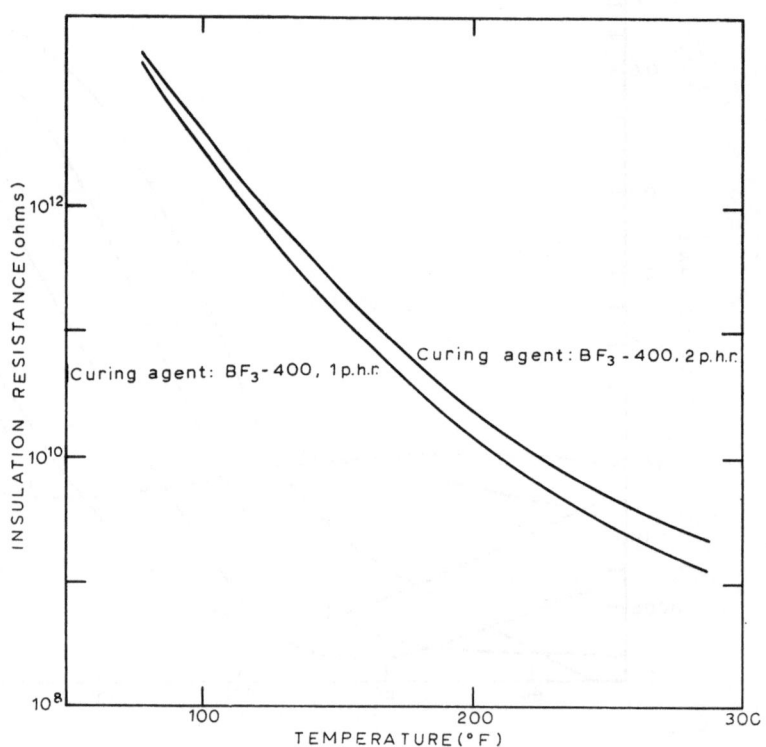

Insulation resistance vs. temperature
of glass cloth-epoxide laminate. Glass,
"E" glass, 12 ply, 181 weave; keying
agent, Volan A; resin, Epikote 1310.

Resin	Test Temperature °F	Dielectric Constant κ'	Dissipation Factor tan δ	Loss Index
Phenolic	75	4.26	0.022	0.0937
	300	4.59	0.039	0.1790
	400	4.67	0.042	0.1961
	500	5.13	0.096	0.4925

"E" glass style 181 fabric with approximately 35 percent resin content. Properties measured at 9.35 GHz.

Type F120 style 181 fabric   Ref. 28

	1 Megacycle	3000 Megacycles
Dielectric Constant, Dry	5.92	4.23
Dielectric Constant, Wet	5.92	---
Dissipation Factor, Dry	0.021	0.083
Dissipation Factor, Wet	0.024	---

Volume Resistivity, Dry (ohm-cms)	2.05×10^{13}
Volume Resistivity, Wet (ohm-cms)	5.55×10^{12}
Surface Resistivity, Dry (ohms)	4.60×10^{13}
Surface Resistivity, Wet (ohms)	5.55×10^{12}

Resin	Test Temperature °F	Dielectric Constant κ'	Dissipation Factor tan δ	Loss Index
Polybenzi midazole	75	4.02	0.0120	0.0482
	300	3.94	0.0080	0.0315
	400	4.08	0.0096	0.0392
	500	4.20	0.0087	0.0365

E glass style 181 fabric. Resin content approximately 19 percent.
Test frequency 9.35 GHz.

Ref. 61

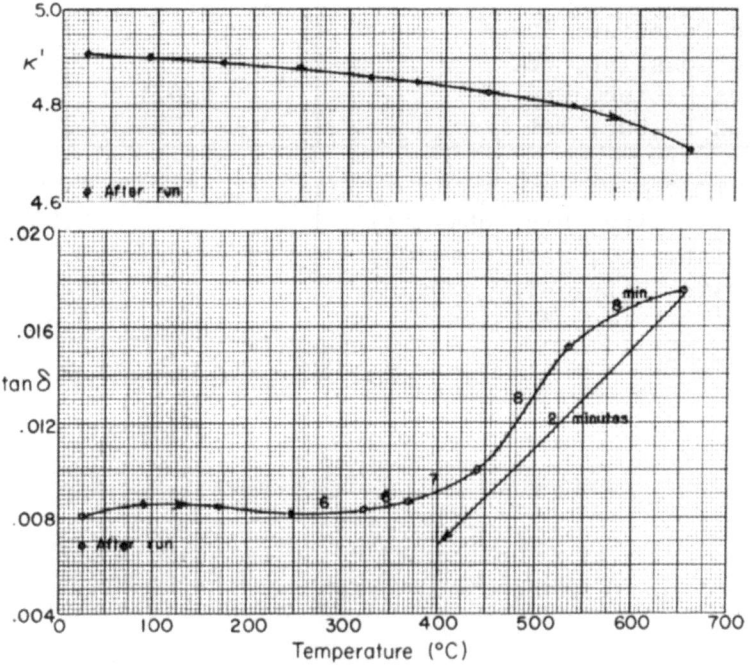

Resin content approximately 24 percent.
Laminate density 1.949 g/cm^3.

S glass style 181 with HTS finish
E glass style 181 with A-1100 finish

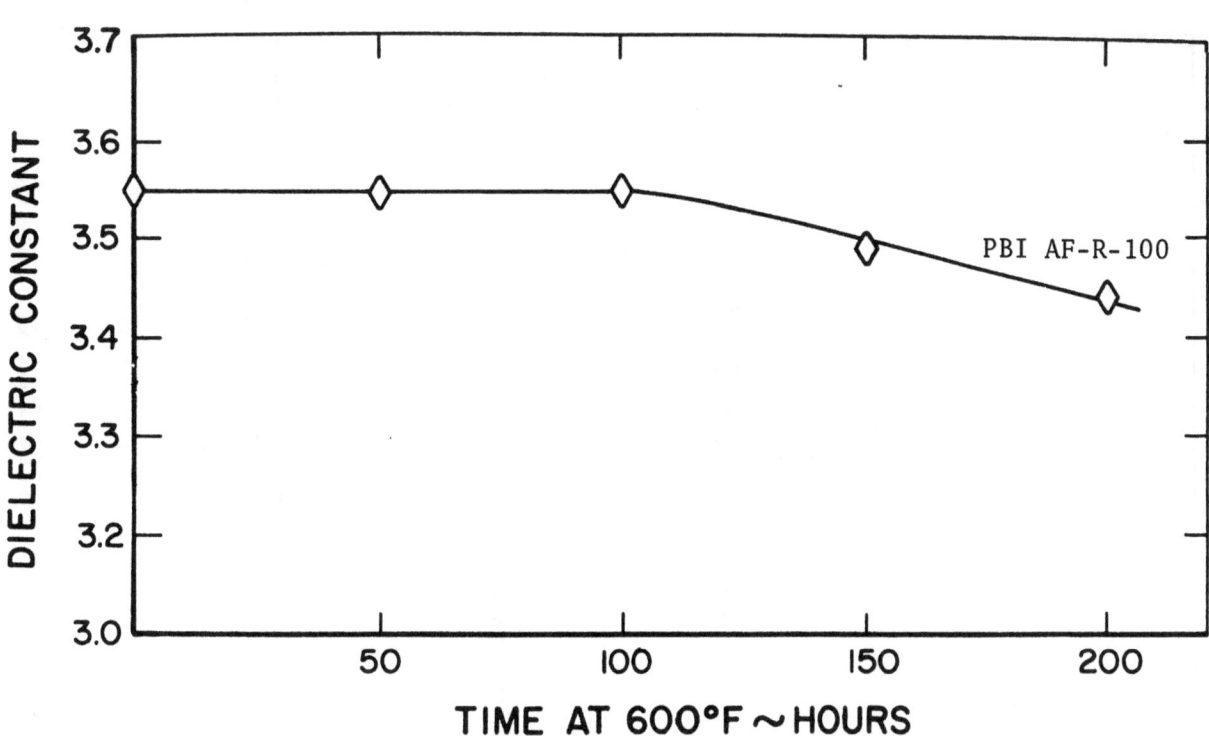

Effect of 600° F Aging in Air on Dielectric Constant of PBI
at 9.375 KMc

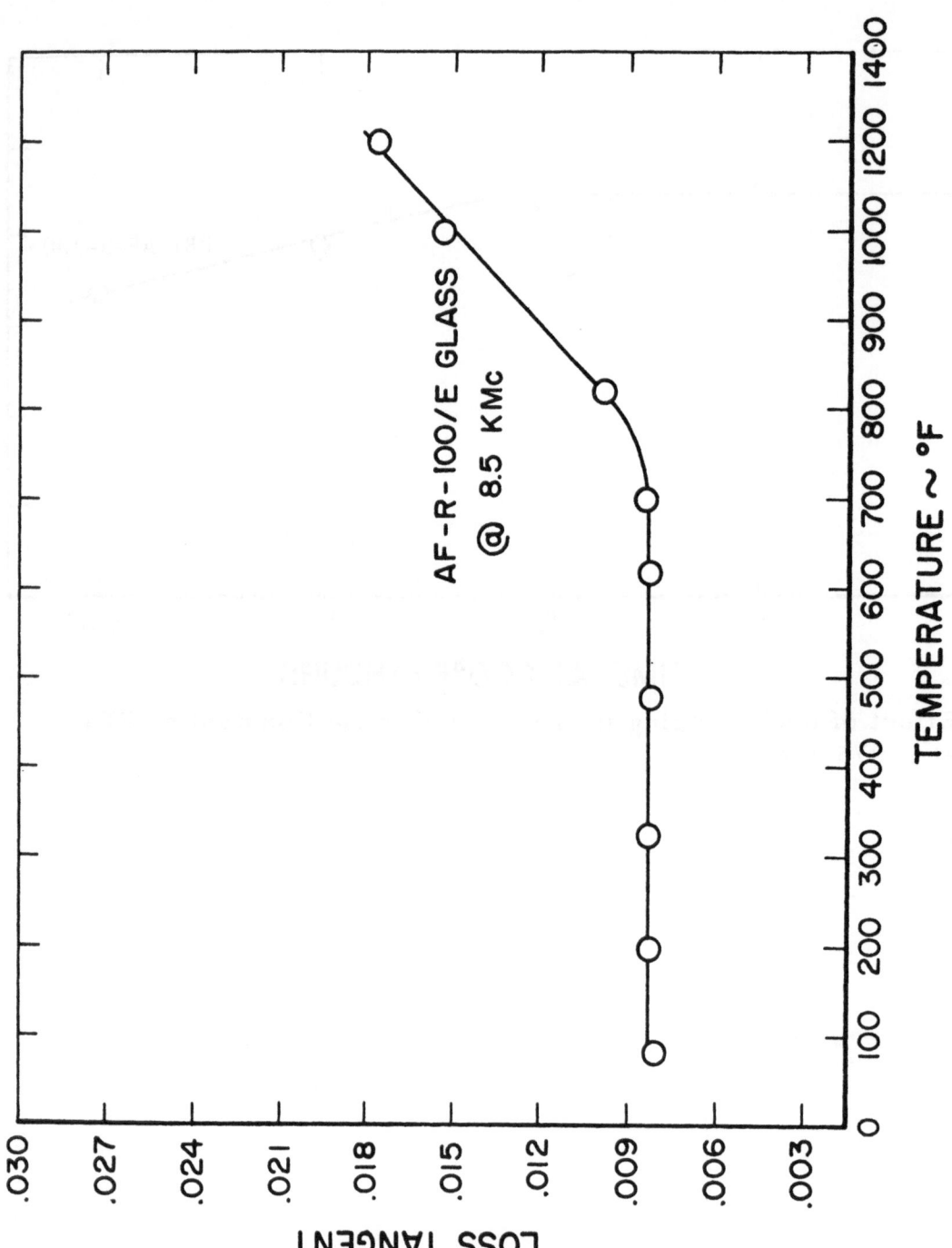

Effect of Temperature on Loss Tangent of AF-R-100 PBI Composite

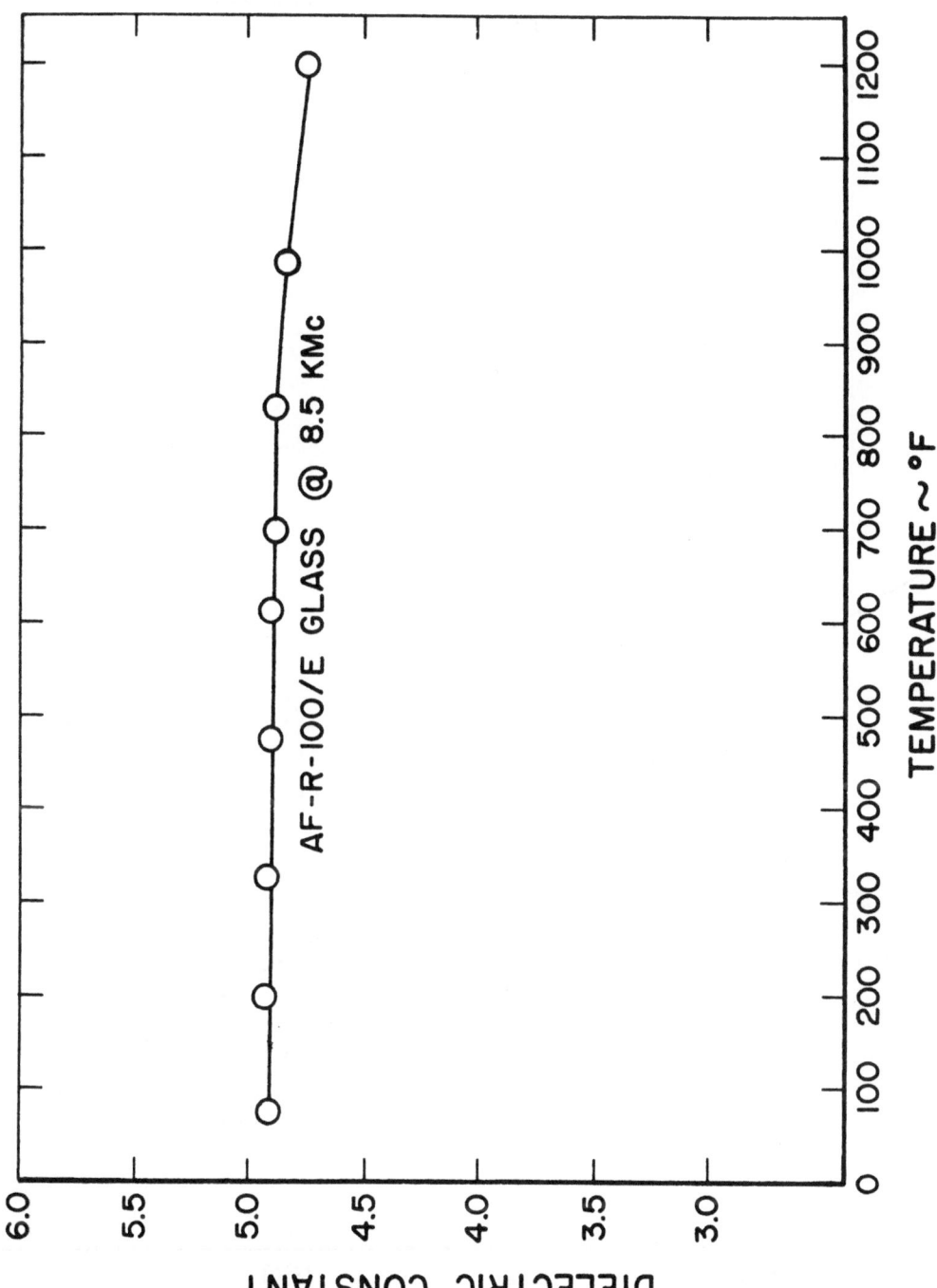

Effect of Temperature on Dielectric Constant of AF-R-100 PBI

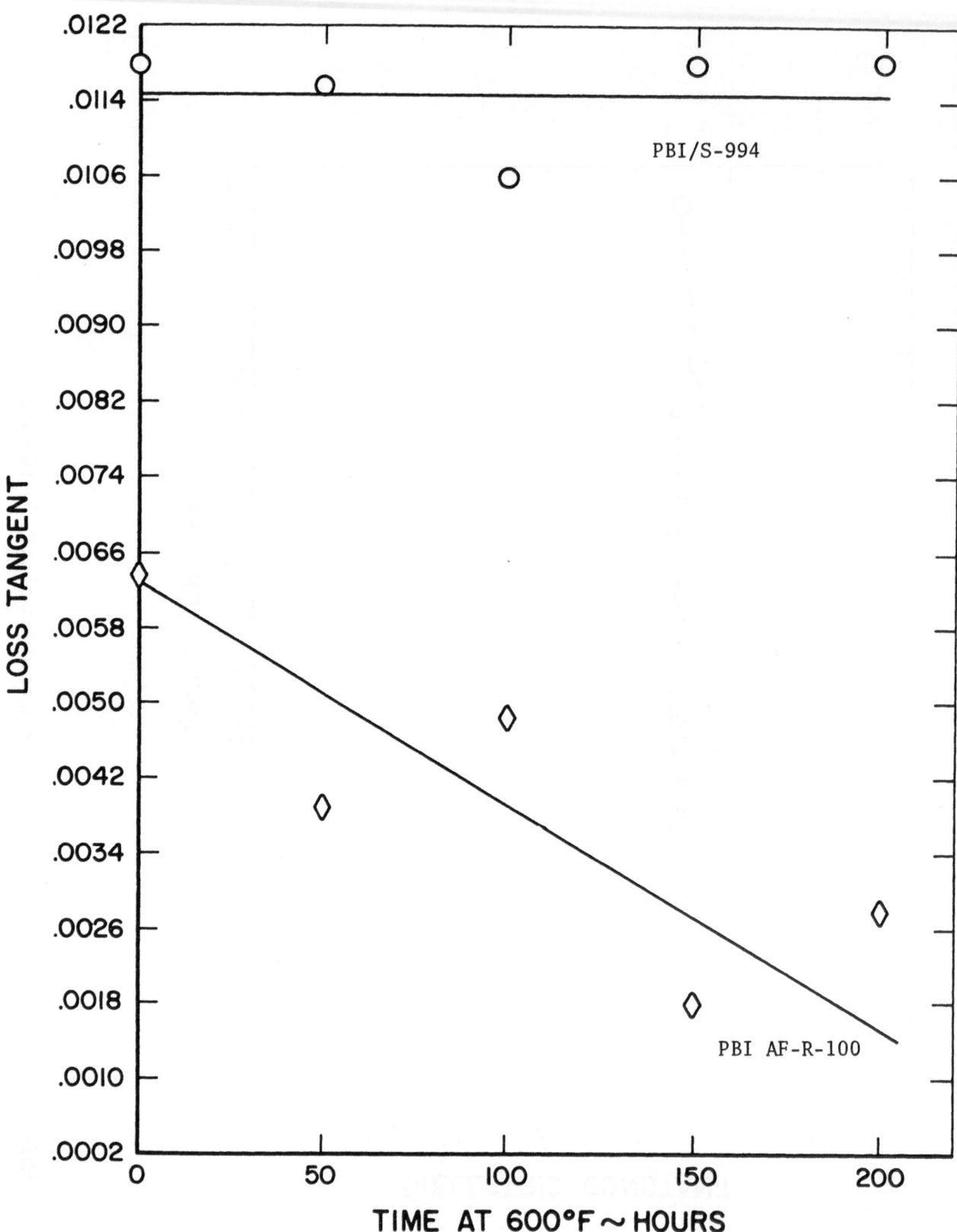

Effect of 600° F Aging in Air on Loss Tangent of PBI at 9. 375 KMc

E glass style 181 with A-1100 finish

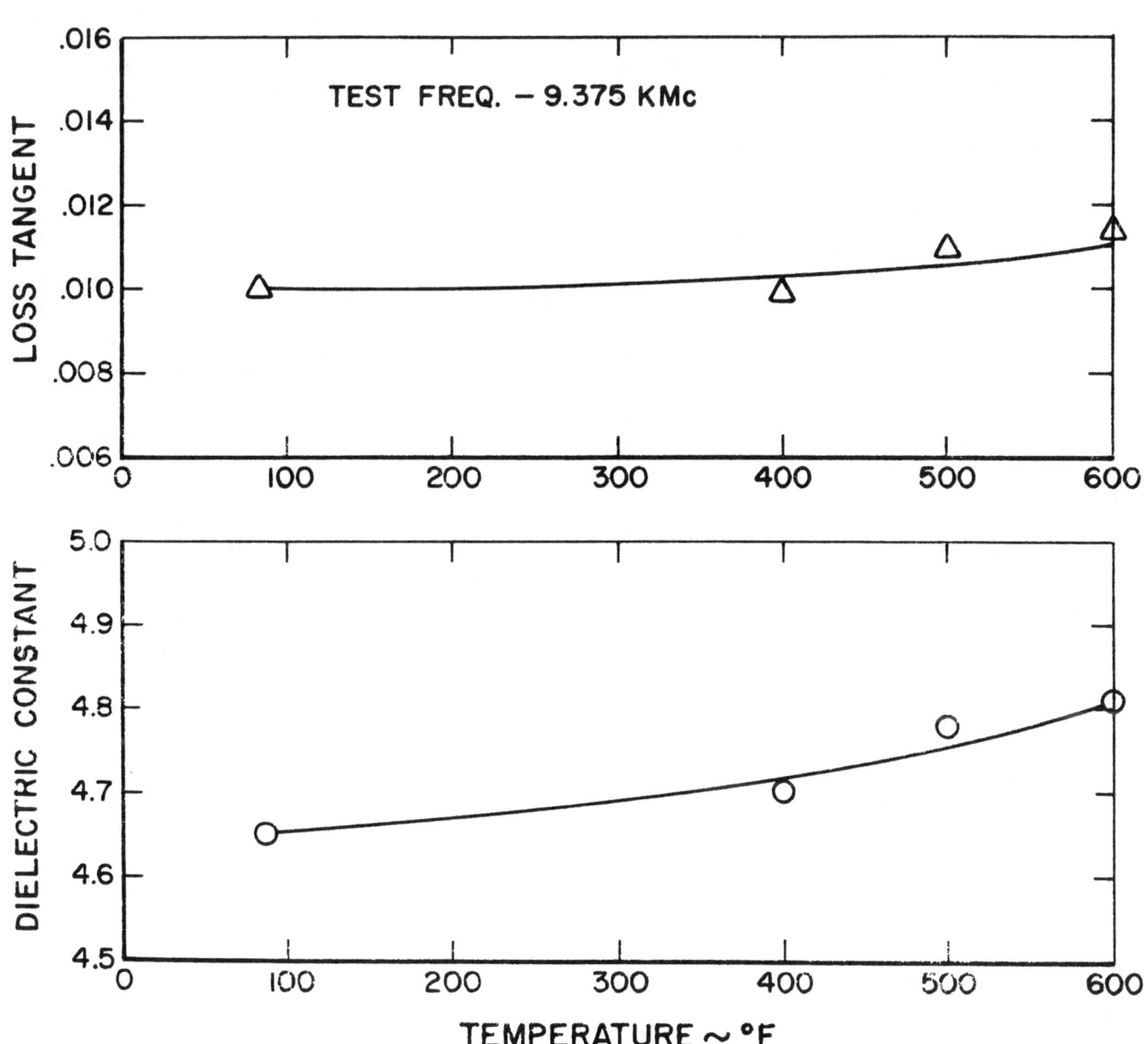

Effect of Temperature on Dielectric Properties of Polybenzothiazole
Laminates

FIBERGLASS POLYESTER

Composition is mineral filled polyester resin reinforced
with random fiberglass mat.

Property	Test Method	Units	Glastic TSF (25)*	Glastic UTS Flame Retard (25)*	Glastic Rod Stock (24)*
Dielectric Strength, ST Condition A	ASTM D 229	V/mil	400	350	150
Dielectric Strength, ST After 200 hrs at 200°C	ASTM D 229	V/mil	250	250	150 (150°C)
Parallel Dielectric Strength, SS Condition A D48/50	ASTM D 229	kV	40 15	40 15	10 -
Power Factor 60 cycles	ASTM D 150	%	3.0	2.5	5.0
Dielectric Constant 60 cycles	ASTM D 150				5.0
Arc Resistance Condition A	ASTM D 495	seconds	120	120	120
Arc Tracking Resistance	ASTM D 2302	minutes	120	25	-
Thermal Conductivity	ASTM C 177	Btu/hr/ft^2/in/°F	-	-	2.0
Thermal Expansion Coefficient	ASTM D 690	in/in°C	-	-	2×10^{-5}

*Numbers in parenthesis refer to items listed in the References.

Resin	Test Temperature °F	Dielectric Constant κ'	Dissipation Factor tan δ	Loss Index
Polyester 1	75	4.15	0.012	0.0498
	300	4.23	0.021	0.0888
	400	4.36	0.025	0.1090
	500	4.40	0.036	0.1584
Polyester 2	75	4.68	0.011	0.0515
	300	4.73	0.025	0.1183
	400	4.70	0.035	0.1645
	500	5.01		
TAC polyester	75	4.35	0.015	0.0653
	300	4.37	0.017	0.0743
	400	4.41	0.016	0.0706
	500	4.46	0.018	0.0803

"E" glass style 181 fabric with approximately 35 percent resin content. Properties measured at 9.35 GHz.

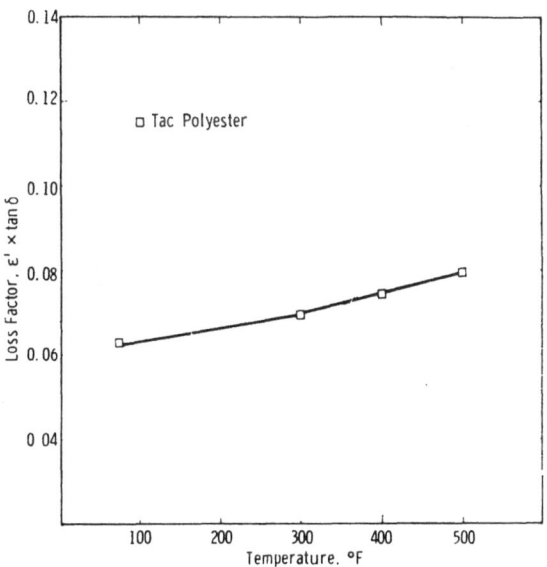

Loss factor at 9.35 GHz of style-120 glass-fabric laminates as function of temperature (40 percent resin content).

49

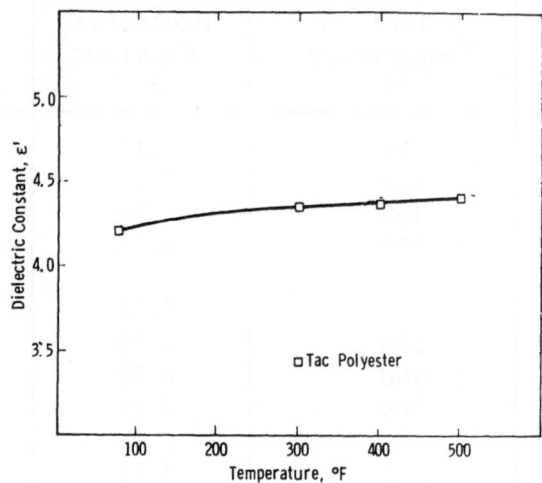

Dielectric constant at 9.35 GHz of
style-120 glass-fabric laminates as
function of temperature (40 percent
resin content).

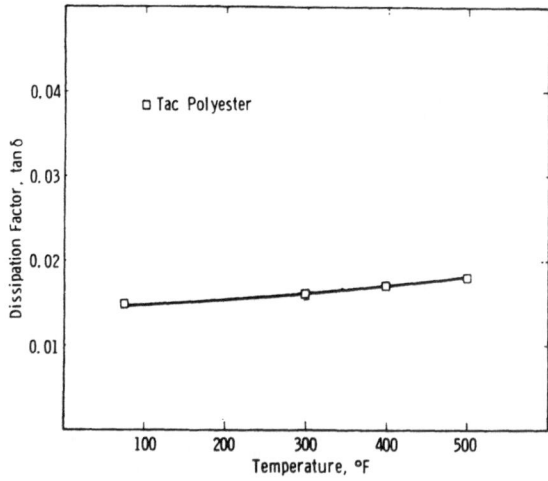

Dissipation factor at 9.35 GHz of
style-120 glass-fabric laminates as
function of temperature (40 percent
resin content).

FIBERGLASS-POLYESTER (CONT'D)

The fiberglass used is style 181 fabric.

Property	Test Method	F141 (28)	F149 (28)	Poly-Preg P 603 (60)	Poly-Preg 646 (60)
Dielectric Constant 9.36 GHz	MIL-R-7575				
Dry		3.99	4.25	-	4.31
Wet		-	4.33	-	-
Dielectric Constant 10^6 Hz	FSTM 406				
Dry		2.4		4.27	
Wet		2.5		-	
Dielectric Constant 10^3 Hz	FSTM 406	-	-	4.48	
Dissipation Factor 9.36 GHz	MIL-R-7575				
Dry		0.0106	0.011	-	0.016
Wet		-	0.015	-	-
Dissipation Factor 10^6 Hz	FSTM 406				
Dry		0.0096	-	0.012	
Wet		0.0120		-	
Dissipation Factor 10^3 Hz	FSTM 406	-	-	0.008	

Effect of temperature on electrical properties of
diallylisophthalate (DAIP) modified polyester resin (1)
and laminate (3). At 8,500 - 10,000 megacycles (2).

Temperature °C	Dielectric Constant	Loss Tangent
20	4.49	.014
37	4.60	.015
87	4.69	.018
137	4.74	.018
187	4.70	.018
237	4.72	.018
260	4.75	.020
20	4.47	.011

(1) - California Chemical Company, Chevron 6100 - 85 pbw/DAIP
      - 15 pbw.

(2) - Tested according to MIL-R-25042A

(3) - Laminate from pretreated 181-150 Garan finish cloth, 12 ply,
      1/8 inch laminate, 70% glass.

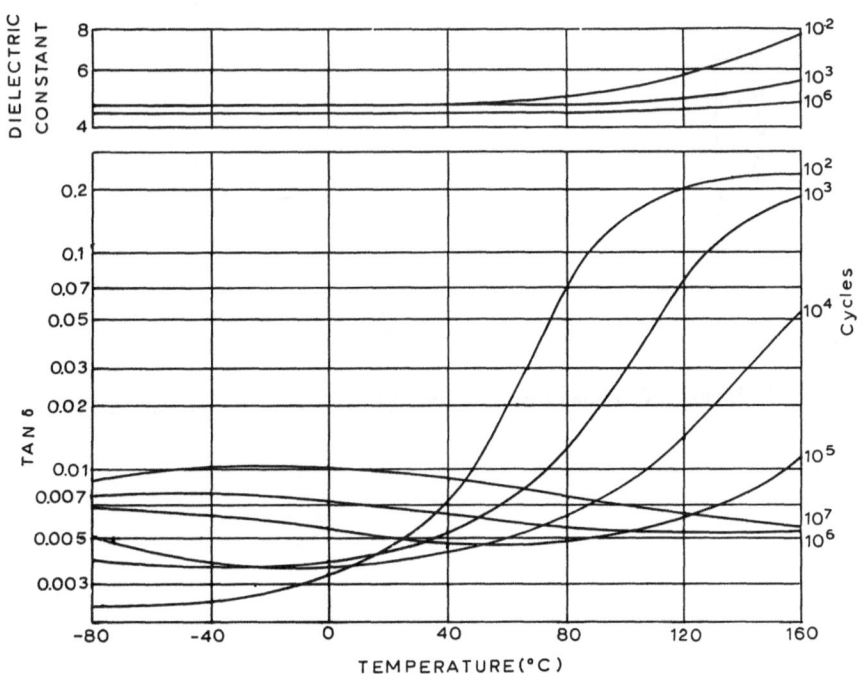

Dielectric constant and loss angle vs.
temperature at various frequencies. Glass
chopped-strand-mat-polyester laminate.

Ref. 16

Filled Glass Mat Laminates

	As Molded (2.5 min at 240°F) (0.5% USP-245)	Post Cured (16 hrs at 135°C) (2 hrs at 180°C) (2 hrs at 200°C)	Aged (10 days at 220°C)
Electrical Properties *			
Arc Resistance, sec	170-180	175-185	185-195
Dielectric Strength			
Perpendicular, S.T., volts/mil	470-480	485-495	495-505
Perpendicular, S.S., volts/mil	450-460	460-470	470-480
Parallel, S.T., kv	65-70	60-65	55-60
Parallel, S.S., kv	60-65	55-60	50-55
Dielectric Strength, wet			
Perpendicular, S.T., volts/mil			485-495
Perpendicular, S.S., volts/mil			470-480
Parallel, S.T., kv			55-60
Parallel, S.S., kv			50-55
Power Factor, 100 cycles, %	1.0-1.2	0.85-1.05	1.4-1.6
Dielectric Constant	4.1-4.3	4.1-4.3	4.1-4.3
Volume Resistivity, ohm-cm	1.45×10^{16}	1.40×10^{16}	0.40×10^{16}
Surface Resistivity, ohm	$> 1.0 \times 10^{14}$	$> 1.0 \times 10^{14}$	$> 1.0 \times 10^{14}$
Insulation Resistance, ohm	$> 1.0 \times 10^{13}$	$> 1.0 \times 10^{13}$	$> 1.0 \times 10^{13}$
Track Resist., inclined plane, min	650-700	750-850	

* 33% Koplac V7000-15, 33% glass, 34% ASP-400.

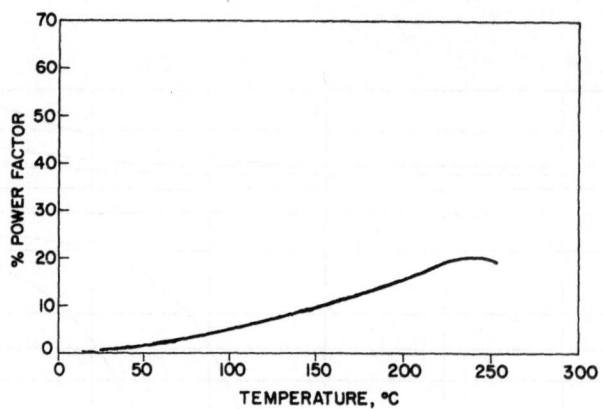

Influence of temperature on the % power factor of V7000-15

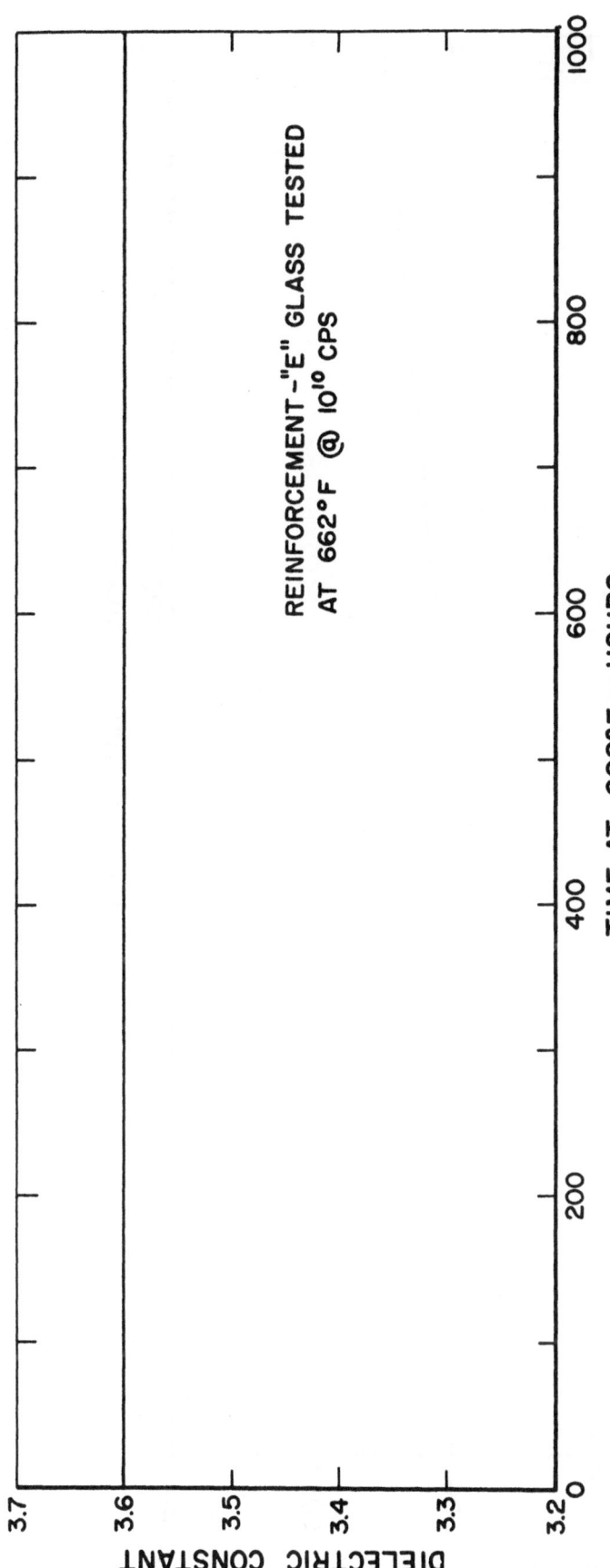

Effect of 600^0 F Aging on Dielectric Constant of AF-R-5000 Polyimide Laminates

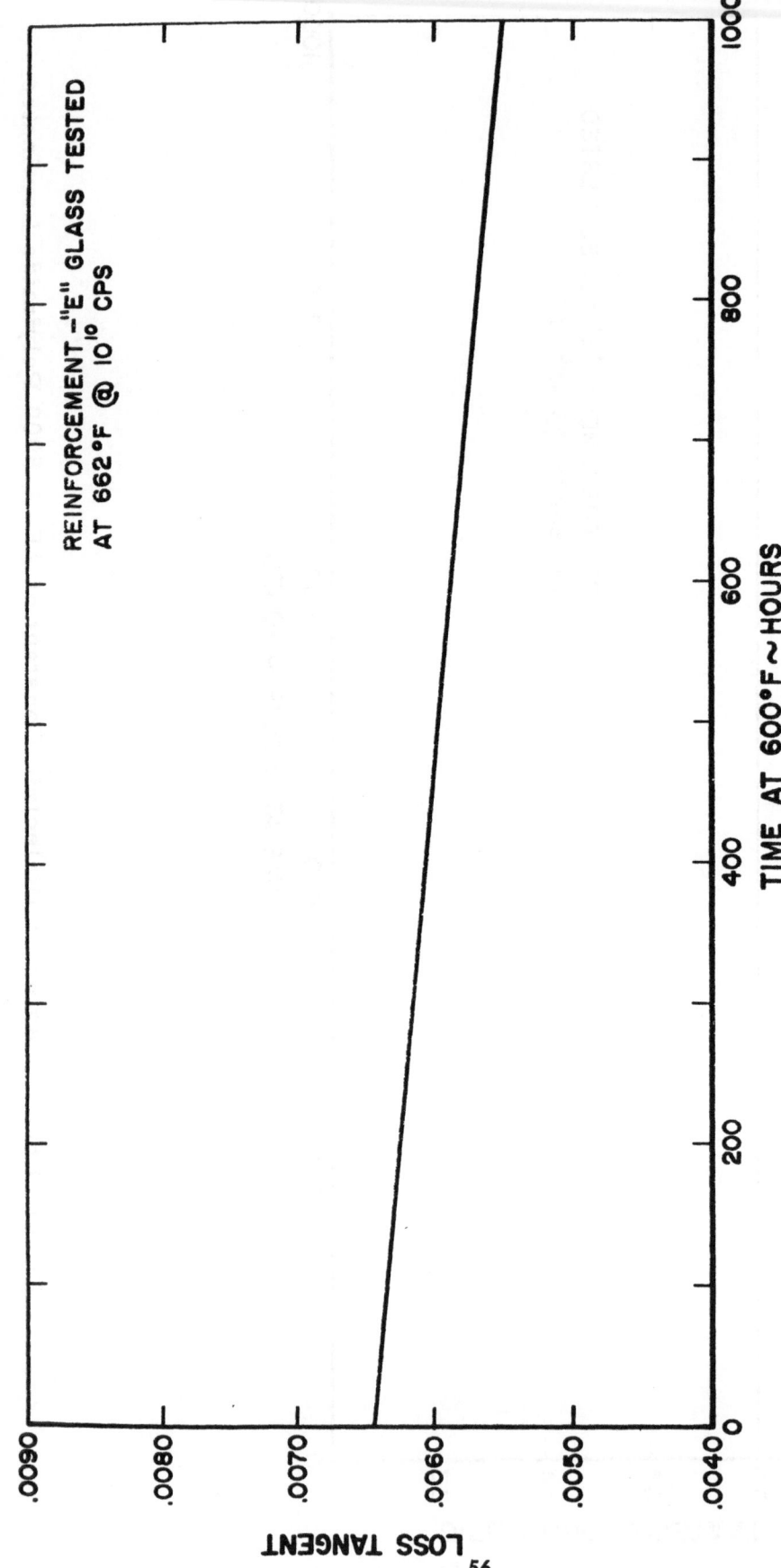

Effect of 600⁰ F Aging on Loss Tangent of AF-R-5000 Polyimide Laminates

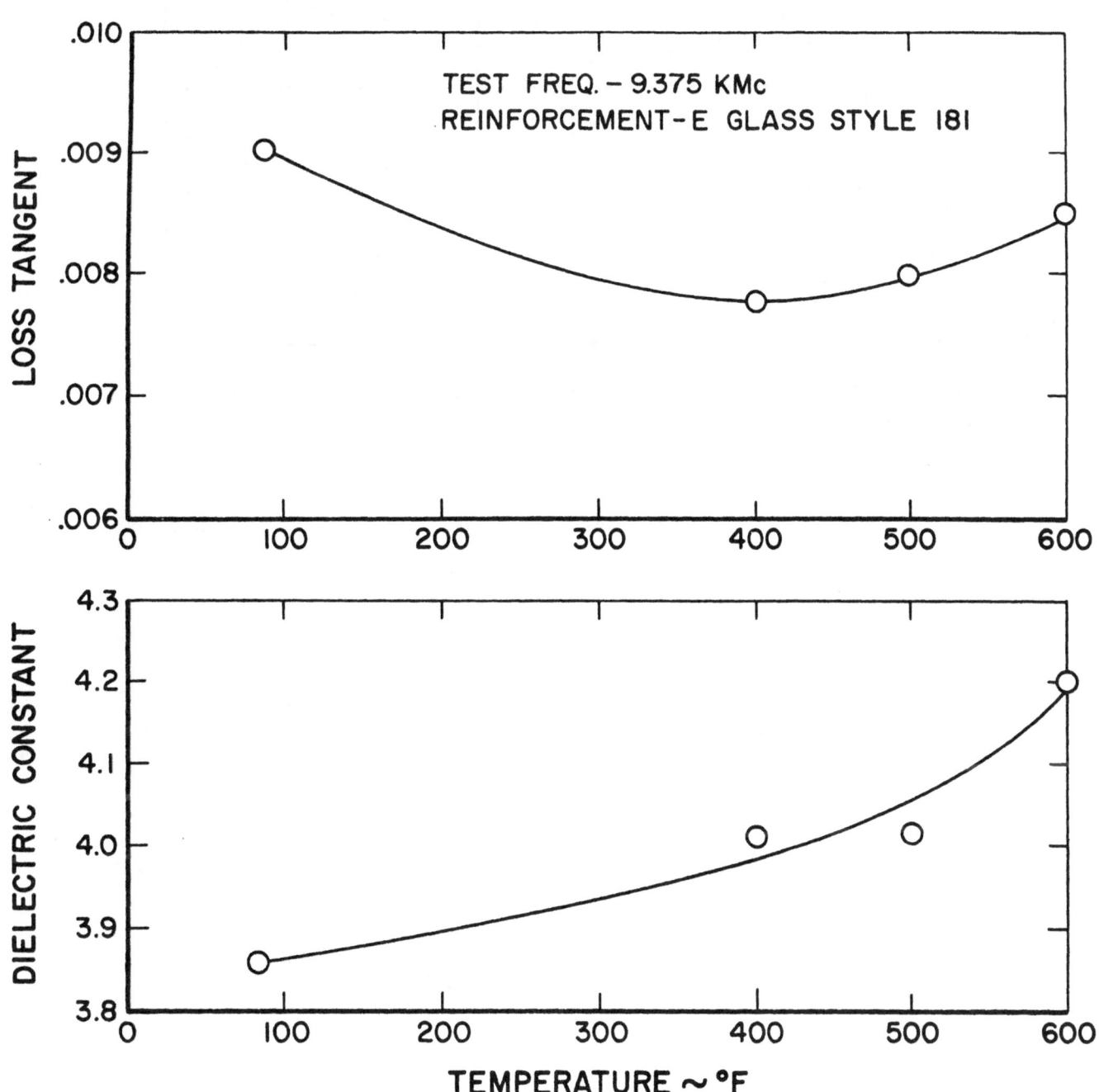

Effect of Temperature on Dielectric Properties of Skybond 700 Polyimide Laminates

Resin	Test Temperature °F	Dielectric Constant, κ'	Dissipation Factor tan δ	Loss Index
Polyimide 1	75	3.81	0.0075	0.0286
	300	3.88	0.0117	0.0454
	400	3.89	0.0075	0.0292
	500	4.01	0.0111	0.0445
Polyimide 2	75	4.20	0.0071	0.0298
	300	4.30	0.0082	0.0353
	400	4.34	0.0110	0.0477
	500	4.34	0.0150	0.0651

E glass style 181 fabric with approximately 35 percent resin content. Properties measured at 9.35 GHz.

Glass Fabric   Ref. 61

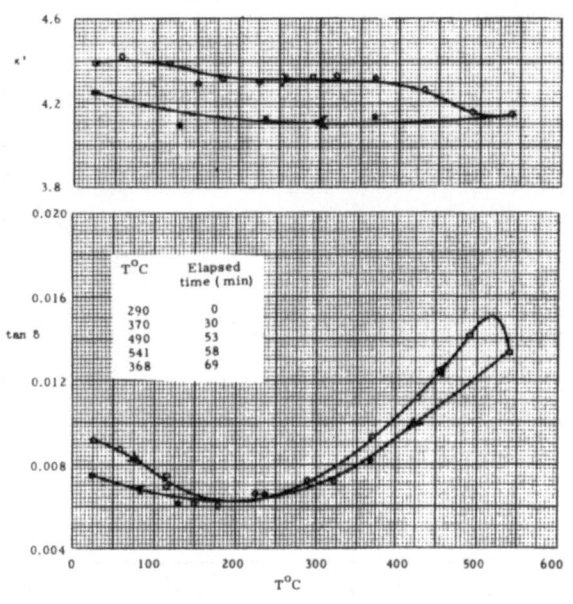

Fiberglass laminate with 181 glass cloth and a polyol cross-linked polyimide resin, 8.52 GHz

Fiberglass laminate with 181 glass cloth and a polyimide resin, 8.52 GHz

Skybond 700 Laminates [*]

Property	As is	D 24/23[*]	D 48/50[*]	C96/35/90[*]
Dielectric strength				
Short time parallel to laminate (volts)	55,000	--	32,000	--
Step-by-step parallel to laminate (volts)	38,000	--	16,000	--
Short time (volts/mil)	179	--	--	--
Stepwise (volts/mil)	140	--	--	--
Dielectric Constant (1MC)	4.10	4.30	4.81	--
Dissipation Factor (1MC)	.00445	.00639	.01650	--
Insulation resistance (megohms)	1.9×10^7	--	--	1.4×10^2
Volume Resistivity (ohm-cms)	2.47×10^{15}	--	--	1.16×10^{11}
Surface Resistivity (ohms)	3.35×10^{14}	--	--	2.90×10^{10}

* ASTM conditioning procedures,   ASTM D618-61

X-Band Data (8.5 KMC)

Temperature	Dielectric Constant	Dissipation Factor
Room Temperature	3.74	0.016
50°C	3.74	0.015
100°C	3.74	0.014
150°C	3.74	0.018
200°C	3.74	0.013
250°C	3.74	0.010
300°C	3.70	0.015

* E glass 181 style fabric with A-1100 finish.

Property	Test Method	Units	F170* (28)	F171* (28)	Pyralin** (11)	Pyralin*** (11)
Dielectric Constant 8.5 GHz	U		3.74	3.76	3.9	3.9
Dielectric Constant 10^6 Hz	U		4.1			
Dissipation Factor 8.5 GHz	U		0.016	0.016	0.015	0.014
Dissipation Factor 10^6 Hz	U		0.0044			
Thermal Conductivity	U	cal/cm sec°C			4.36×10^{-4}	
Thermal Expansion Coefficient -70° to 90°C 90° to 170°C 170° to 400°C	U	in/in/°C			0.74×10^{-5} 0.34×10^{-5} 0.69×10^{-5}	

  * E glass style 7781 fabric
  ** E glass style 181 fabric with A-1100 finish
*** S-994 glass roving

Resin	Test Temperature °F	Dielectric Constant κ'	Dissipation Factor tan δ	Loss Index
Silicone	75	3.98	0.0081	0.0322
	300	4.17	0.0069	0.0288
	400	4.31	0.0066	0.0284
	500	4.46	0.0070	0.0312

"E" glass style 181 fabric with approximately 35 percent resin content properties measured at 9.35 GHz.

Property	Test Method	Units	F130* (28)	F131* (28)	F130** (46)	Poly-Preg 5860* (60)
Dielectric Constant 9.36 GHz Dry	MIL-R-25506		3.946		3.60	4.17
Dielectric Constant 9.36 GHz Wet	MIL-R-25506		4.021			
Dielectric Constant 10^6 Hz Dry	MIL-R-25506		4.03	3.9		
Dielectric Constant 10^6 Hz Wet	MIL-R-25506		4.12	4.0		
Dissipation Factor 9.36 GHz Dry	MIL-R-25506		0.0082		0.002	0.013
Dissipation Factor 9.36 GHz Wet	MIL-R-25506		0.0106			
Dissipation Factor 10^6 Hz Dry	MIL-R-25506		0.0019	0.002		
Dissipation Factor 10^6 Hz Wet	MIL-R-25506		0.0075	0.010		
Dielectric Strength 9.37 GHz	U	V/mil	100			
Arc Resistance 9.37 GHz	U	seconds	244			

* E glass
** D glass

RT/duroid 5870 is a polytetrafluoroethylene
laminate reinforced with randomly oriented microglass fibers.

Property	ASTM Method	Test Values
Dielectric Strength, Short Time, volts/mil.	D149-55T	300
Dielectric Constant, 1 MHz	D1531-58T	2.35
Dissipation Factor, 1 MHz	D1531-58T	0.0005
Dielectric Constant, 10 GHz	MIL-P-13949	2.35
Dissipation Factor, 10 GHz	MIL-P-13949	.0012
Surface Resistivity, Ohms	D257-57T	
As Received		3.0×10^{14}
96 hours, 100% R.H., 23°C		3.0×10^{14}
Volume Resistivity, Ohm - Cm		
As Received		2.0×10^{13}
96 hours, 100% R.H., 23°C		2.0×10^{13}
Arc Resistance	D495-56T	No track up to melting at 180 sec.
Thermal Expansion Coefficient x 10^{-5}	U	
Longitudinal Direction, 0-100°F		1.6
Transverse Direction, 0-100°F		4.0
Thickness Direction, 0-100°F		10.0
Longitudinal Direction, 100°F-350°F		1.0
Transverse Direction, 100°F-350°F		2.0
Thickness Direction, 100°F-350°F		10.1
Thermal Conductivity, $\dfrac{\text{BTU - in.}}{\text{Hr. -Sq. Ft. -°F}}$	U	1.8

Dielectric Constant vs. Frequency (Hz)

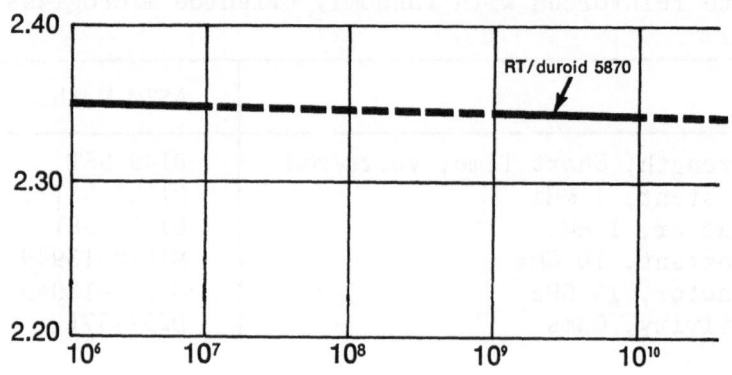

Dissipation Factor vs. Frequency (Hz)

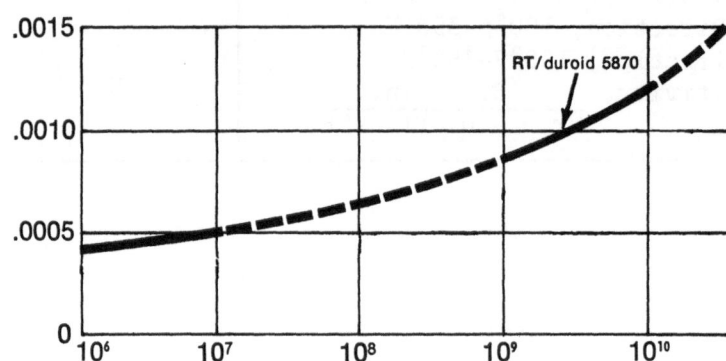

Fluorglass E 650/2-1200

Freq., GHz	T^{o}C	E ⊥		E ∥	
		κ	tan δ	κ	tan δ
8.5	23	2.505	.0014	2.847	.0036
4	-195	2.533	.00082	2.896	.00172

DiClad-522

E ⊥ sheet   All values of tan δ multiplied by 10^4

T^{o}C	Freq. (Hz)	10	10^2	10^3	10^4	10^5	10^6	10^7	5.5×10^7	9×10^7	3.14×10^9*
25	κ	2.739	2.740	2.738	2.737	2.735	2.734	2.733	2.732	2.731	2.712
	1 tan δ	8.6	7.0	6.7	6.1	6.3	6.95	7.7	10.0	11.7	22.5
100	κ		2.710	2.705	2.704	2.698	2.696	2.683			2.680
	1 tan δ		11.1	8.10	8.25	7.17	7.07	7.7			31
250	κ		2.554	2.534	2.522	2.503	2.502	2.49			
	1 tan δ		79.0	36.3	20.35	14.9	11.6	10.6			
-78	κ		2.796	2.793	2.790	2.784	2.78	2.78			2.752
	1 tan δ		4.2	5.9	6.8	7.1	7.7	9.8			17
-195	κ	2.801	2.799	2.794	2.792	2.787					2.758
	1 tan δ	.0005	2.2	4.5	5.1	5.4					12
-269	κ	2.789	2.789	2.784	2.783	2.780					
	1 tan δ	.0003	1.2	2.0	2.2	2.1					

* Copper cavity

E ∥ sheet

T^{o}C	Freq. (Hz)	3×10^8	10^9	3×10^9	8.5×10^9	1.4×10^{10}	2.4×10^{10}
25	κ	3.155	3.153	3.152	3.146	3.133	3.127
	tan δ	28	30	33	40	48	52
100	κ				3.11		
	tan δ				39		
250	κ				3.03		
	tan δ				36		
-54	κ				3.17	3.13	
	tan δ				35	39	
-195	κ				3.22	3.12	
	tan δ				28	31	

Experimental Laminates with a Reemay Reinforcement

	Units	Cond.	Epoxy Matrix	XXXP Phenolic Matrix	X Phenolic Matrix	Poly-ester Matrix
Dielectric Constant (1 mc)	–	A	3.2–3.4	3.9	4.0	3.1
	–	D24/23	3.3–3.6	4.0	4.0	3.1
Dielectric Str. Perp. St	vpm	A	917			
Par. SXS kv	kv	D48/50	80–110	75	90	70
Insulation Resistance	m ohm	C-96/35/90	10^9–$10''$	10^9		10^9
Surface Resistance	m ohm	C-96/35/90	10^8–$10''$	10^8	10^9	10^{11}

Dielectric Properties of PRD-49-III Composites at 9.3 GHz Frequency

Sample	Fiber	v/o Fiber	Resin	Temp. °F	Condition	Fiber Orientation	Dielectric Constant	Loss Tangent
P5289-18A	49-III (120 Style)	58	BP-907	77	---	$\perp$ to $\overline{E}$	3.25	0.00995
"	"	"	"	"	---	$\parallel$ to $\overline{E}$	3.67	0.0127
"	"	"	"	315	---	$\parallel$ to $\overline{E}$	4.05	0.0509
"	"	"	"	77	After 315 °F measurement	$\parallel$ to $\overline{E}$	3.68	0.0167
"	"	"	"	77	After 24 hrs in H_2O	$\parallel$ to $\overline{E}$	3.65	0.0186
P5289-25	49-III (120 Style)	54	Epon-828	77	---	$\parallel$ to $\overline{E}$	3.45	0.0159
"	"	"	"	315	---	$\parallel$ to $\overline{E}$	3.88	0.0406

Thermal Conductivity of PRD-49-III/Epoxy Composites[*]

Fiber Form	v/o Fiber	Heat Flux Direction	Thermal Conductivity Btu/hr/ft^2/°F/ft
120 style fabric	46	Across fabric layers	0.123
120 style fabric	46	Parallel to wrap	0.525
181 style fabric	59.5	Across fabric layers	0.122
Unidirectional	54.0	Transverse to fibers	0.087
Unidirectional	54.0	Parallel to fibers	1.01

* Test methods unknown.

9.3 GHz FREQUENCY

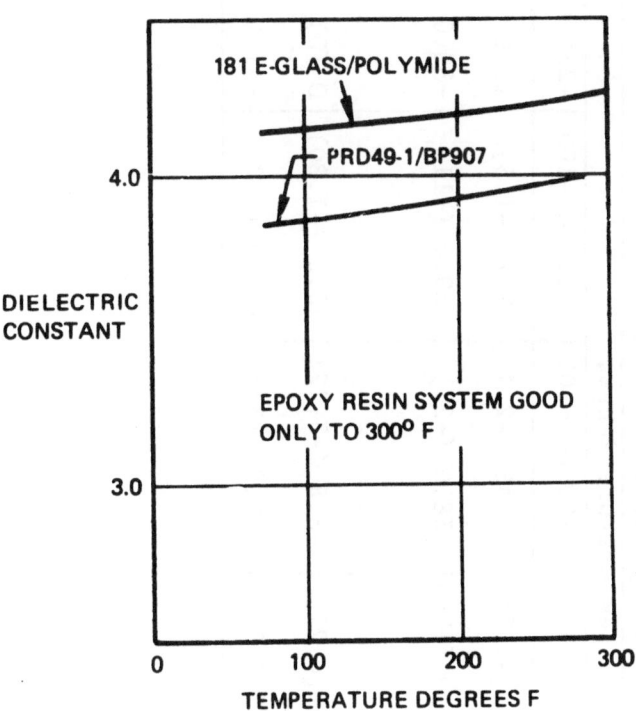

EPOXY RESIN SYSTEM GOOD
ONLY TO 300° F

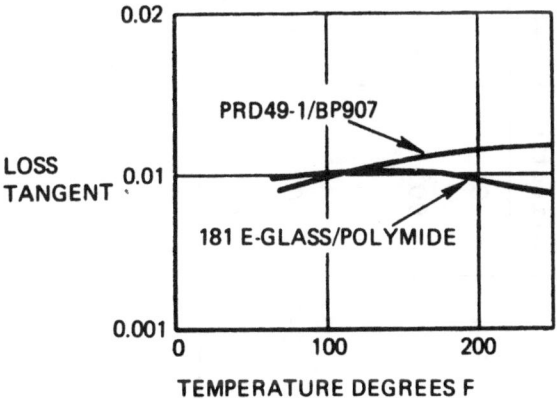

Test methods in reference.

Coefficient of Thermal Expansion for Unidirectional PRD 49-1 Fiber Reinforced/Epoxy Composites
Unidirectional Fiber Orientation

Composite	Test Direction △	Coefficient of Thermal Expansion, 10^{-6} in/in/°C (10^{-6} in/in/°F) Temperature Range, °C					
		-250° to +150	+100 to +150	-100 to +100	-200 to -100	-250 to -200
ERLA 4617/PRD 49-1	0°	-3.68 (-2.0)	-6.6 (-3.7)	-4.45 (-2.47)	-2.7 (-1.5)	+1.0 (+0.56)
	90°	+45.0 (+25)	+101.0 (+56)	+38.0 (+21)	+20.0 (+11)	+30.0 (+17)
Resin System No. 2/ PRD 49-1	0°	-4.95 (2.75)	-7.6 (4.22)	-5.8 (-3.22)	-3.2 (-1.80)	0 (0)
	90°	+61.0 (+34)	+93.0 (+52)	+56.0 (+31)	+41.0 (+23)	+41.0 (+23)

△ 0 = Longitudinal measurement, parallel with fiber
   90° = Transverse measurement, perpendicular with fiber

9.3 GHz FREQUENCY *

RESIN	V_f %	TEMP. (°F)	DIELECTRIC CONSTANT	LOSS TANGENT
POLYIMIDE (P13N)	48	70	3.40	0.00645
		100	3.42	0.00727
		200	3.47	0.00792
		225	3.46	0.00775
		275	3.42	0.00726
		300	3.39	0.00717
		400	3.40	0.00973
		500	3.45	0.0148
		600	3.52	0.0210
	36	70	3.28	0.00505
		100	3.29	0.00555
		200	3.31	0.00621
		250	3.32	0.00654
		300	3.30	0.00660
		400	3.31	0.00889
		500	3.39	0.0132
		600	3.45	0.0194

* Test methods given in reference

Resin	Reinforce-ment	Dielectric Constant, κ'	Dissipation Factor, tan δ	Loss Index
Epoxy novolac 1	120 glass 503 quartz	4.55 3.46	0.019 0.015	0.0865 0.0519

Quartz style 503 fabric. Test frequency 9.35 GHz. Test methods unknown.

Property Fabric Description	Epoxy Epon 828/CL
Oz/sq yd	8.4
Weave	8H satin
Finish	A1100
Dielectric constant X-band, dry-room temp (9,375 megacycles)	3.470
Loss tangent X-band, dry-room temp	0.0092

Ref. 46

QUARTZ-PHENOLIC   Ref. 46

Property Fabric Description	Phenolic V-204
Oz/sq yd	8.4
Weave	8H satin
Finish	A1100
Dielectric constant X-band, dry-room temp (9,375 megacycles)	3.86
Loss tangent X-band, dry-room temp	0.040

Property Fabric Description	PBI AF-R-100
Oz/sq yd	8.4
Weave	8H satin
Finish	A1100
Dielectric constant	
X-band, dry-room temp (9,375 megacycles)	3.360
Loss tangent	
X-band, dry room temp	0.0034

Quartz style 581 with amino silane finish.   Ref. 3

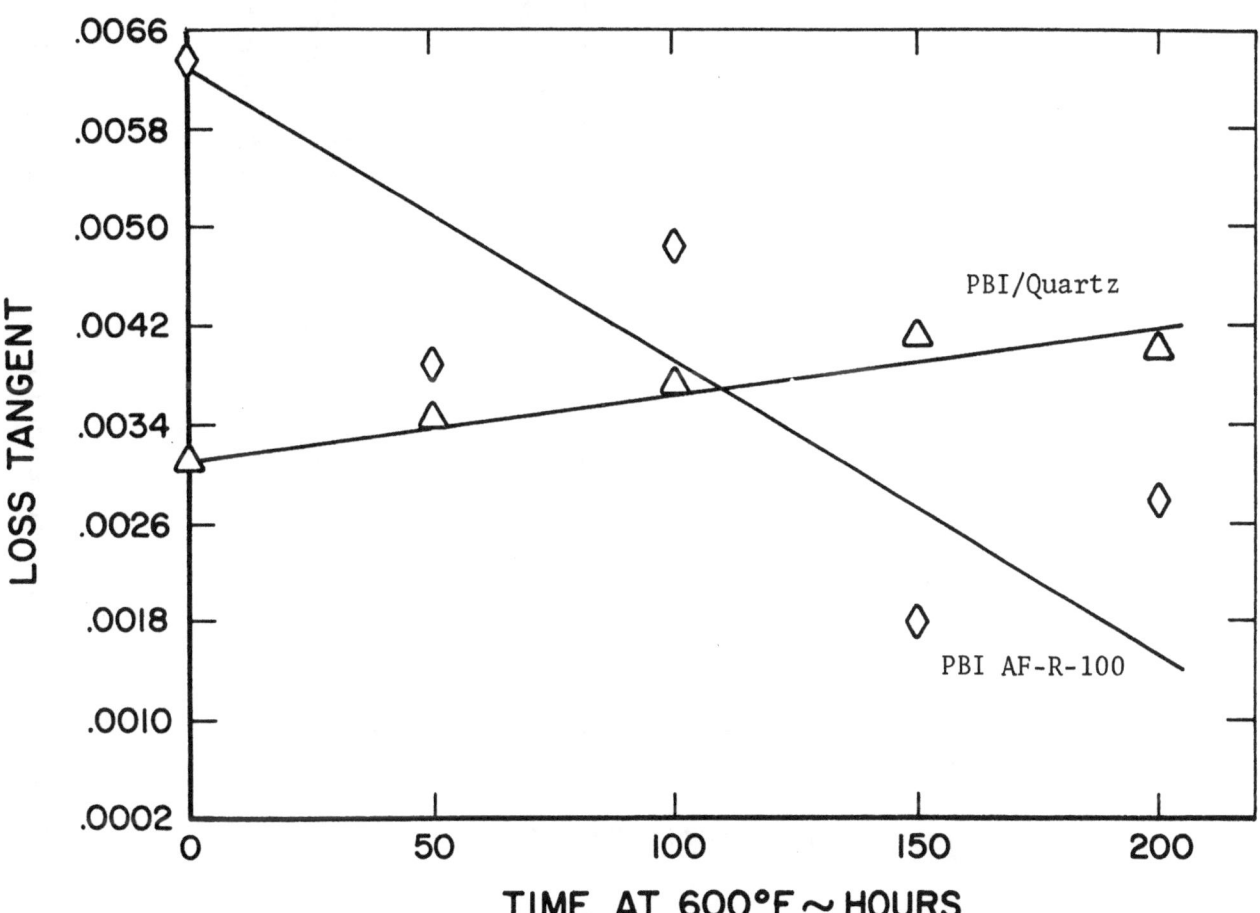

Effect of 600° F Aging in Air on Loss Tangent of PBI at 9.375 KMc

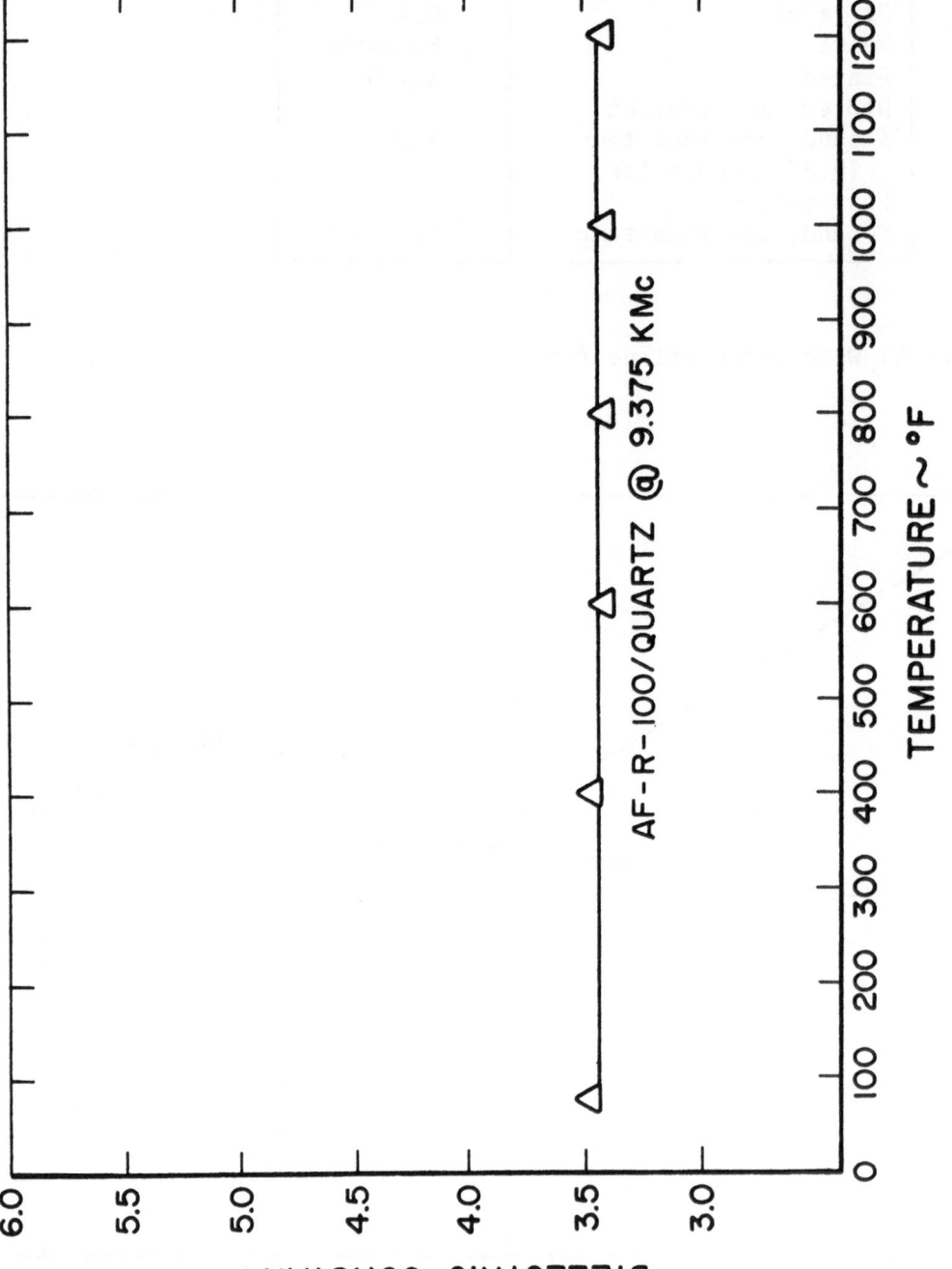

Effect of Temperature on Dielectric Constant of AF-R-100 PBI

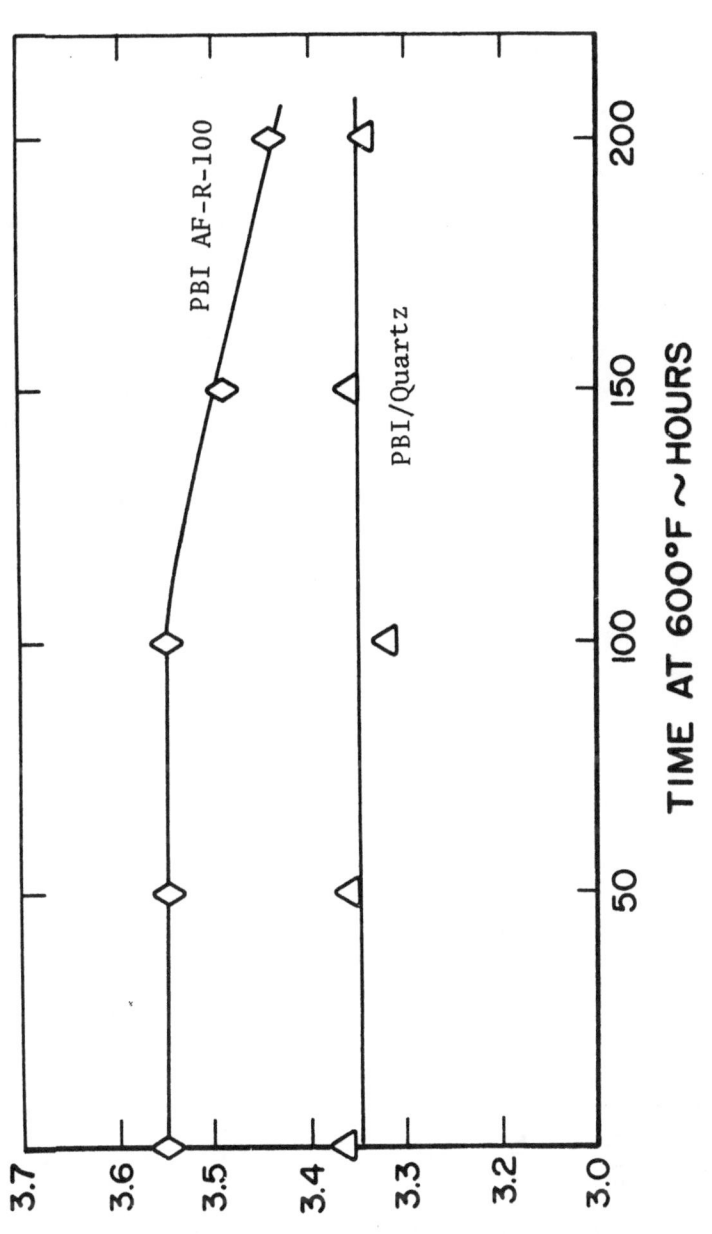

Effect of 600° F Aging in Air on Dielectric Constant of PBI
at 9.375 KMc

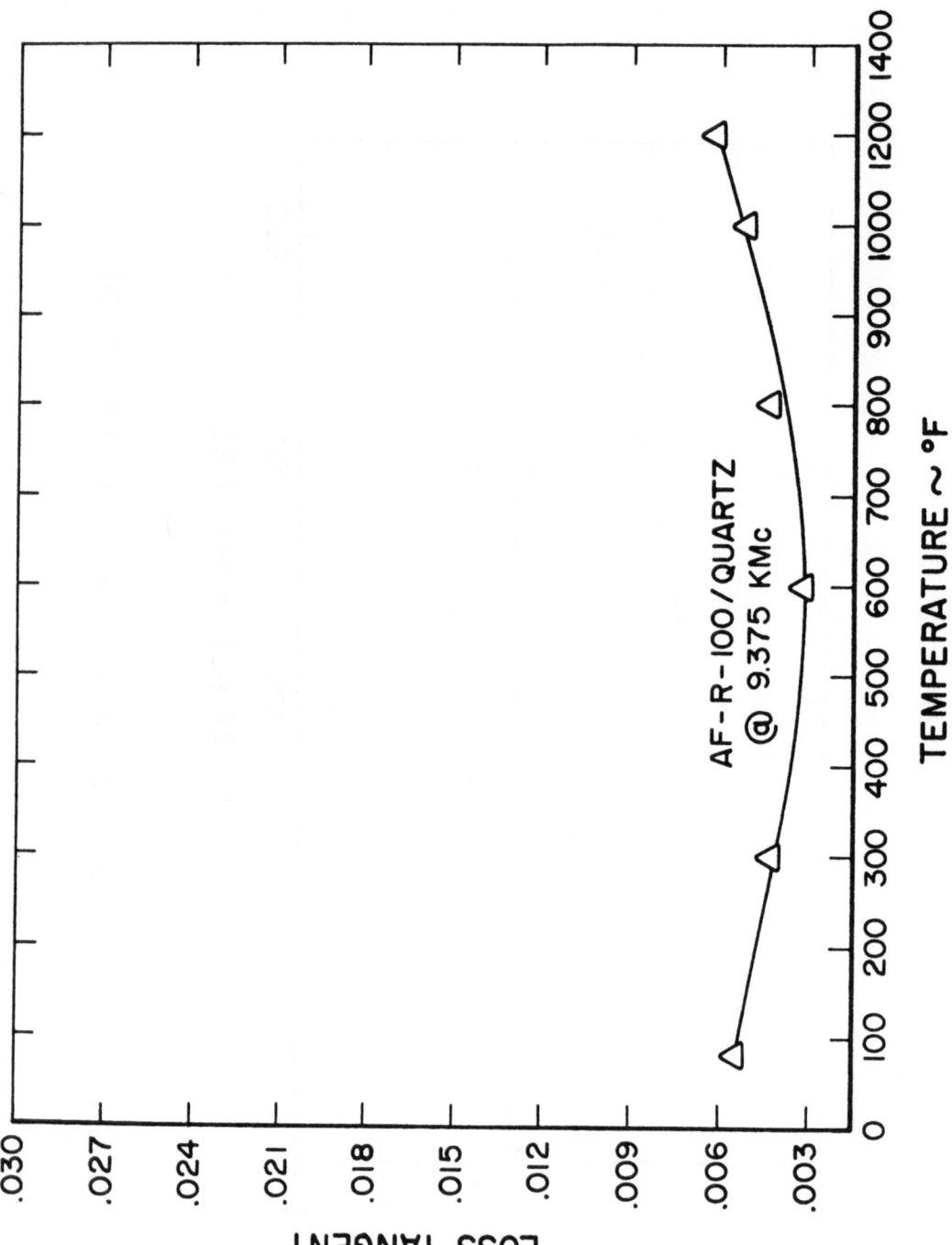

Effect of Temperature on Loss Tangent of AF-R-100 PBI

Resin	Reinforce-ment	Dielectric Constant κ'	Dissipation Factor, tan δ	Loss Index
Polyester 1	503 quartz	3.45	0.005	0.0173
TAC polyester	503 quartz	3.19	0.010	0.0319

* Quartz style 503 fabric. Test frequency 9.35 GHz.

Resin	Reinforce-ment	Dielectric Constant κ'	Dissipation Factor, tan δ	Loss Index
DAIP	503 quartz	3.40	0.005	0.0170

* Quartz style 503 fabric. Test frequency 9.35 GHz.

Resin	Reinforce-ment	Dielectric Constant κ'	Dissipation Factor, tan δ	Loss Index
Diolefin	503 quartz	2.81	0.0039	0.0110

* Quartz style 503 fabric. Test frequency 9.35 GHz.

Resin	Reinforce-ment	Dielectric Constant κ'	Dissipation Factor tan δ	Loss Index
Polyimide	503 quartz	3.16	0.0049	0.0155

Quartz style 503 fabric. Test frequency 9.35 GHz.

Quartz style 581 fabric astroquartz. Polyimide PBI-373.   Ref. 2

Laminate Identification	A-730-2	A-714-1
Thickness, in.	.54	.28
Specific Gravity	1.64	1.68
Resin Content, % W/O	35.0	34.8
Void Content, % V/O	5.7	5.5
Dielectric Constant/Loss Tangent at 9.375 GHz: R.T.	3.27/.008	*
400°F	3.17/.003	
500°F	2.21/.003	
600°F	3.00/.004	

*see Figure on next page.

Coefficient of Linear Thermal Expansion of Quartz/ BPI-373
Polyimide Laminate - Wrap Direction

Temp. Range	75°F-400°F	400°F-500°F	500°F-600°F
CTE 10^{-6} in/in°F:			
Spec. 1	2.31	2.50	1.50
2	1.77	3.25	1.75
3	2.23	2.25	1.75
Avg	2.10	2.67	1.67

Laminate #A-714-1.   Test method given in reference.

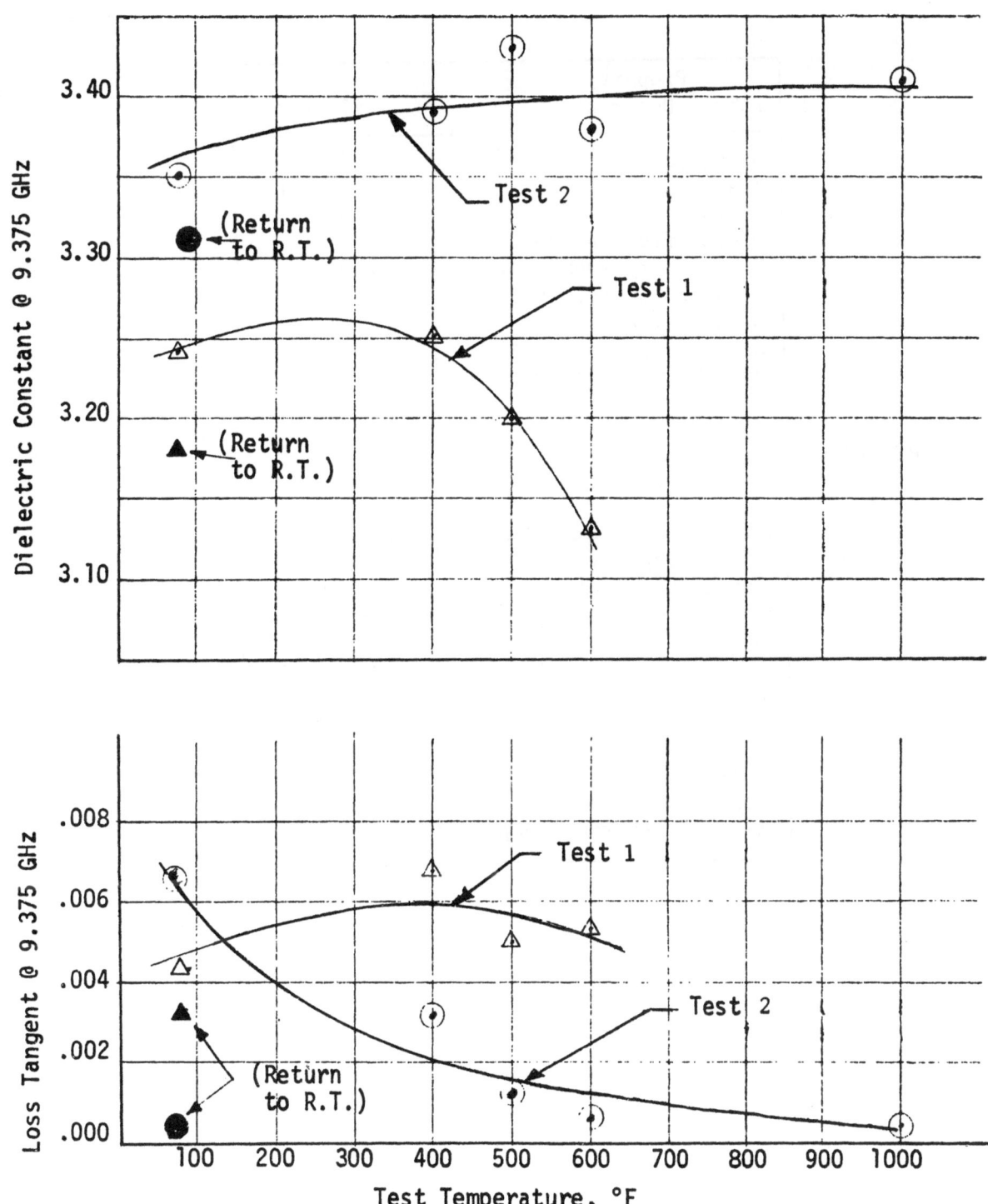

DIELECTRIC PROPERTIES OF QUARTZ/BPI-373 POLYIMIDE
LAMINATE VS. TEMPERATURE

Property Fabric Description	Silicone
Oz/sq yd	8.4
Weave	8H satin
Finish	None
Dielectric constant	
X-band, dry-room temp (9,375 megacycles)	2.93
Loss tangent	
X-band, dry-room temp	0.00098

QUARTZ-TEFLON

Property Fabric Description	Teflon
Oz/sq yd	8.4
Weave	8H satin
Finish	Teflon
Dielectric constant	
X-band, dry-room temp (9,375 megacycles)	2.47
Loss tangent	
X-band, dry-room temp	0.0007

UNIDIRECTIONALLY SOLIDIFIED EUTECTICS   Ref. 20

Thermoelectric Properties of the
Controlled Sb-InSb and Te-Bi$_2$Te$_3$ Eutectics

Material	ρ ohm-cm	Q μv/o	K w/cm^o	
Sb	4.4×10^{-5}	35	0.18	
InSb	10^{-3}	-325	0.16	
InSb-Sb eutectic				
with 4.3 rods	3.05×10^{-4}	-18	0.11	‖ Sb rods
	33.0×10^{-4}	-71	0.08	⊥ Sb rods
with 8.5 rods	3.0×10^{-4}	-12	0.12	⊓ Sb rods
	33.0×10^{-4}	-53	0.09	⊥ Sb rods
with 28 rods	3.15×10^{-4}	-8.2	0.14	⊓ Sb rods
	34.0×10^{-4}	-28	0.10	⊥ Sb rods
Bi$_2$Te$_3$	10^{-3}	214	0.02	
Te	4.0×10^{-1}			
Bi$_2$Te$_3$-Te eutectic				
	5.4×10^{-4}	18 (20 C)		‖ growth di-rection
		22 (40 C)		
		25 (60 C)		
	2.2×10^{-3}	45 (20 C)		⊥ growth di-rection
		55 (40 C)		
		65 (60 C)		

DISPERSION-STRENGTHENED NICKEL AND ALUMINUM   Ref. 62

Physical Properties of Dispersion-Strengthened
Nickel and Aluminum at Room Temperature

Alloy	Dispersoid	Density	Thermal Conductivity at 70° F. Btu (ft-hr-°F)$^{-1}$	Electrical Resistivity (microhm-cm)
TD Nickel[a]	2% ThO$_2$	0.322	50.0	7.6
High Purity Nickel[a]	None	0.320	54.5	7.0
TD NiC[b]	2% ThO$_2$	0.306	8.3	108.0
Aluminum[c]	None	0.097	120.0	2.9
SAP[c]	13% Al$_2$O$_3$	0.100	100.0	4.2
M257[c]	8% Al$_2$O$_3$	0.098	110.0	3.4
M486[d]	Intermetallic Compound	0.104	60.0	6.7

[a] <0.1% total impurities other than thoria.
[b] 20.0% Cr added to Ni matrix.
[c] <1.0% total impurities other than alumina.
[d] Al-7.8% Fe-1% (Cr-V-Ti-Zr).

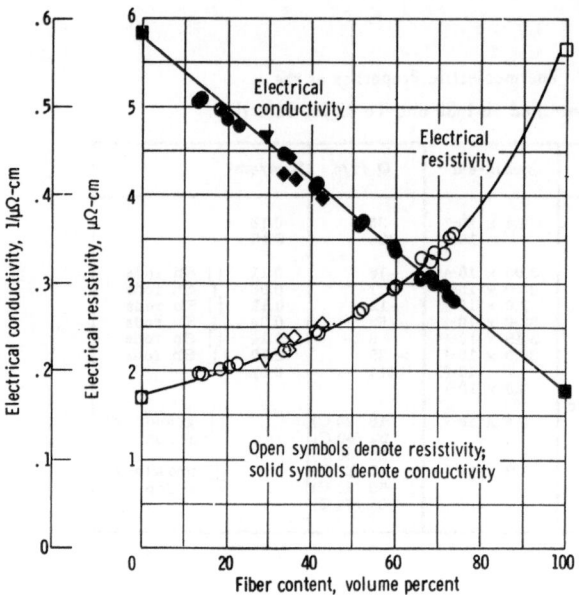

Electrical resistivity and conductivity of tungsten-fiber-reinforced copper composites.

ELECTRICAL RESISTIVITY

Glass Fiber: "E" - .0006" Dia. -Parallel Oriented
Coating Alloy: 2% Zn 0.2% Cd 1100 Al
Matrix Alloy: 2014 Al
Vacuum Injection Casting
Heat Treatment: Solutioned and Aged

Percent Glass By Volume	Resistivity (microhm-cm)	Av. Resistivity (microhm-cm)	Calculated Resistivity (micron-cm)
0		3.13	3.13
15	6.42		
15	6.45		
15	6.54	6.49	3.69
15	6.43		
15	6.64		
19	6.66		
19	6.30		
19	6.29	6.47	3.86
19	6.64		
19	6.48		
22.5	6.50		
22.5	6.55	6.525	4.00
30	6.50		
30	6.45	6.42*	4.47
30	6.30		
30	5.71	5.74**	4.47

 * Samples air cooled immediately after casting
** Held at 900°F for three hours, then slow cooled in furnace

THERMAL EXPANSION DATA

30-40% Glass in a 2014 Aluminum Matrix
Glass Fiber: "E" - .0006" Dia. -Parallel Oriented
Coating Alloy: 2% Zn 0.2% Cd 1100 Al
Vacuum Injection Casting
As Cast

	Temperature °F	$\frac{\Delta L}{L}$
First Cycle	75	0
	161	.0004
	300	.0011
	502	.0020
	715	.0024
	1030	--
	1054	.0023
	407	.0004
	75	-.0008
Second Cycle	75	0
	990	.0029
	75	0
Third Cycle	70	0
	207	.0005
	357	.0013
	484	.0020
	688	.0023
	960	.0027
	764	.0020
	574	.0018
	413	.0012
	227	.0005
	70	--
Fourth Cycle	70	0
	303	.0013
	670	.0022
	942	.0024
	562	.0017
	222	.0003
	70	-.0002

MOLDED COMPOSITE

Random Distribution, Cut Fibers

Polyester Premix Molding Compound Reinforced with Crocidolite Asbestos
(Short View Fibers)

Dielectric Constant:	
@ 1 KHz	7.3
@ 10 KHz	6.1
Powder Factor:	
@ 1 KHz	0.22
@ 10 KHz	0.014
Volume Resistivity:	
@ 500 Vdc, ohm-cm	1.2×10^{12}

Random Distribution, Cut Fibers

Property		Test Method FTMS 406
Arc Resistance Secs	125	4011
Dielectric Strength (S/S) Volts/Mil D48/50	> 397	4031
Dielectric Constant 1 KC Dry	3.7	4021
Dielectric Constant 1 KC D24/23	3.8	4021
Dielectric Constant 1 MC Dry	3.5	4021
Dielectric Constant 1 MC D24/23	3.5	4021
Dissipation Factor 1 KC Dry	0.014	4021
Dissipation Factor 1 KC D24/23	0.015	4021
Dissipation Factor 1 MC Dry	0.016	4021
Dissipation Factor 1 MC D24/23	0.020	4021
Volume Resistance Megohms C 720/70/100+Dew	0.06×10^6	MIL-M-14 F

FIBERGLASS-DIALLYL PHTHALATE (DAP) AND ISOPHTHALATE (DAIP)

Random Distribution, Cut Fibers

Property	Test Method	Units	DAP		DAIP	
			1D30 (53) and Poly-Dap 6130 (60)	1D30*F (53) and Poly-Dap 6130F (60)	1M30 (53) and Poly-Dap 6230 (60)	1M30F (53) and Poly-Dap 6230F (60)
Dielectric Constant 10^6 Hz	FTMS 406- 4021					
Dry			4.2	4.2-3.9	4.0	4.3-4.0
Wet			4.3	4.2-4.0	4.0	4.3-4.1
Dielectric Constant 10^3 Hz	FTMS 406- 4021					
Dry			4.4	4.3-4.1	4.2	4.3-4.1
Wet			4.4	4.3-4.1	4.2	4.3
Dissipation Factor 10^6 Hz	FTMS 406- 4021					
Dry			0.013-0.010	0.010-0.017	0.012	0.014-0.011
Wet			0.015-0.012	0.014-0.017	0.014	0.015
Dissipation Factor 10^3 Hz	FTMS 406- 4021					
Dry			0.008	0.007-0.009	0.009	0.008
Wet			0.008-0.010	0.007-0.017	0.008-0.011	0.008-0.011
Dielectric Strength, SS Wet		V/mil	38-420	397-424	391-425	389-410
Electrical Resistance (Volume) Humidity + Dew	MIL-P 19833	10^{12} ohms	0.17	0.14	0.08	0.20
Arc Resistance	U	seconds	125	125	125-150	125

* Flame retardant

87

Random Distribution, Cut Short and Long Fibers

Property	Test Method	Units	EM7IO2 (60)	E7111* (18)	E260* (18)	E360* (18)
Dielectric Constant 10^6 Hz Dry Wet	FTMS 406-4021		5.1 5.1	5.1 -	5.8 6.0 (170°C)	4.5 4.8 (170°C)
Dielectric Constant 10^3 Hz Dry Wet	FTMS 406-4021		5.4 5.5			
Dissipation Factor 10^6 Hz Dry Wet	FTMS 406-4021		0.019 0.021		0.017 0.02 (170°C)	0.0015 0.02 (170°C)
Dissipation Factor 10^3 Hz Dry Wet	FTMS 406-4021		0.013 0.016			
Dielectric Strength, ST Dry Wet	U	V/mil		380 -	420 -	400 -
Dielectric Strength, SS Dry Wet	FTMS 406-4031	V/mil	376 368		390 -	380 -
Electrical Resistivity (Volume)	U	ohm-cm		1×10^{16}		
Arc Resistance	FTMS 406-4011	seconds	186	180		
Thermal Conductivity	U	cal/sec/cm²/°C/cm		10×10^{-4}		
Thermal Expansion Coefficient	U	in/in/°C		1.1×10^{-5} (-30 to +30°C)	0.9×10^{-5} (-30 to +200°C)	1.2×10^{-5} (-30 to +200°C)

Random Distribution, Cut Short and Long Fibers

Property	Test Method	Units	M-2880	M-2037
Dielectric Constant 10^6 Hz	ASTM D150		6.2	8.0
Dissipation Factor 10^6 Hz	ASTM D150		0.020	0.030
Dielectric Strength, ST Dry Wet	ASTM D149	V/mil	300 275	230 150
Dielectric Strength, SS Dry Wet		V/mil	250 250	220 135
Arc Resistance	ASTM D495	seconds	180+	180+
Thermal Conductivity	ASTM C177	cal-cm/sec cm^2°C	10×10^{-4}	
Thermal Expansion Coefficient	ASTM D696	in/in/°C	1.6×10^{-5}	1.6×10^{-5}

Random Distribution, Cut Short Fibers

Property	Test Method	Units	ZYTEL 7010-33 (9)	ZYTEL 7110-33 (9)	NYLAFIL G-1/30 (17)	NYLAFIL G-10/40 (17)
Dielectric Constant 60 Hz	ASTM D150				4.0	4.4
Dielectric Constant 10^3 Hz Dry Wet	ASTM D150		4.52 25.0	4.22 –	3.9 –	4.4 –
Dielectric Constant 10^6 Hz Dry Wet	ASTM D150		3.73 10.7	3.38 –	3.4 –	4.1 –
Dissipation Factor 60 Hz	ASTM D150				0.018	0.009
Dissipation Factor 10^3 Hz	ASTM D150		0.0199	0.0185	0.022	0.011
Dissipation Factor 10^6 Hz	ASTM D150		0.0221	0.0177	0.017	0.018
Dielectric Strength, ST	ASTM D149	V/mil	530	630	500	480
Dielectric Strength, SS	ASTM D149	V/mil	440	510	400	400
Electrical Resistivity (Volume) Dry Wet	ASTM D257	ohm-cm	4.7×10^{15} 2.3×10^9	4.0×10^{14} 3.0×10^9	5.5×10^{15} –	2.6×10^{15} –
Arc Resistance	ASTM D495	seconds			148	100
Thermal Conductivity	Cenco-Fitch	Btu-in/hr ft^2°F			1.5	3.3
Thermal Expansion Coefficient	ASTM D696	in/in/°F	1.3×10^{-5}	1.0×10^{-5}	2.1×10^{-5}	1.4×10^{-5}

Random Distribution, Cut Long and Short Fibers

		FM 4030-190	FM 5064
Dielectric Constant, 1 MC Dry	ASTM D150	5.0	6.0
Dissipation Factor, 1 MC Dry	ASTM D150	0.02	0.025
Dielectric Strength, Volts/Mil, S/T Dry	ASTM D149	380	320
S/S Dry		320	300
S/T Wet		380	260
S/S Wet		300	180
Volume Resistivity, Ohm-cm	ASTM D257	1×10^{13}	1×10^{13}
Coefficient of Linear Expansion, in./in./°C −30°C to +30°C	ASTM D696	1.5×10^{-5}	1.8×10^{-5}
Thermal Conductivity, Cal/Sec/Cm^2/°C/Cm	ASTM C177	10×10^{-4}	9×10^{-4}

S/S, step by step

S/T, short time

Random Distribution, Cut Long Fibers

	PBD-160
Dielectric Constant @ RT, 1 mc	4.2
Dielectric Constant @ 170°C 1 mc	4.25
Dissipation Factor @ RT 1 mc.	.003
Dissipation Factor @ 170°C 1 mc	.0035
Dielectric Strength V/mil S/T	420 +
Dielectric Strength V/mil S/S	420 +
Coefficient of Linear Expansion in/in/°C -30 to +200°C	1.3×10^{-5}

FIBERGLASS-POLYIMIDE   Ref. 18

Random Distribution, Cut Long Fibers

	PI-560
Dielectric Constant @ RT, 1 mc	4.9
Dielectric Constant @ 170°C 1 mc	4.9
Dissipation Factor @ RT 1 mc	.005
Dissipation Factor @ 170°C 1 mc	.0045
Dielectric Strength V/mil S/T	380
Dielectric Strength V/mil S/S	350
Coefficient of Linear Expansion in/in/°C -30 to +200°C	$.6 \times 10^{-5}$
Half Life @ 260°C hrs.	5000

FIBERGLASS-POLYCARBONATE

Random Distribution, Cut Short Fibers

Property	Test Method	Polycarbafil G-50/20 (17)	Polycarbafil G-50/40 (17)	Thermocomp DF-1004 (45)	Thermocomp DF-1004 (45)	Units
Dielectric Constant 60 Hz	ASTM D150	3.7	3.8	3.31	3.66	
Dielectric Constant 10^3 Hz	ASTM D150		3.8	3.31	3.68	
Dielectric Constant 10^6 Hz	ASTM D150	3.3	3.7	3.26	3.61	
Dissipation Factor 60 Hz	ASTM D150	0.006	0.003	0.0008	0.0012	
Dissipation Factor 10^3 Hz	ASTM D150		0.002	0.0011	0.0014	
Dissipation Factor 10^6 Hz	ASTM D150	0.009	0.008	0.0079	0.0072	
Dielectric Strength, ST	ASTM D149	475	482	490	465	V/mil
Dielectric Strength, SS	ASTM D149	435	422			V/mil
Electrical Resistivity (Volume)	ASTM D257	1.5×10^{15}	1.5×10^{15}	6×10^{15}	6×10^{15}	ohm-cm
Arc Resistance	ASTM D495	70	100	120	120	seconds
Thermal Conductivity	Cenco-Fitch	0.522	0.374	2.3	2.7	Btu-in/hr/ft^2°F
Thermal Expansion Coefficient	ASTM D696	1.8×10^{-5}	1.2×10^{-5}	2.1×10^{-5}	2.9×10^{-5}	in/in-°F

Random Distribution, Cut Short and Long Fibers

Property	Test Method	Units	S-642 (18)	S-6300 (18)	S-6400**** (18)	Type Electr. Grade (27)
Dielectric Constant 10^6 Hz Dry Wet	LP406-4021		5.42 5.43	6.1* 7.0	4.95* 5.03	
Dielectric Constant 10^3 Hz Dry Wet	LP406-4021		5.07 4.88			
Dissipation Factor 10^6 Hz Dry Wet	LP406-4021		0.0083 0.0113	0.012* 0.051	0.009* 0.0106	
Dissipation Factor 10^3 Hz Dry Wet	LP406-4021		0.0152 0.0082			
Dielectric Strength, ST Dry Wet	LP406-4031	V/mil	390 410	335** 401	380** 475	300-350** -
Dielectric Strength, SS Dry Wet	LP406-4031	V/mil	290 400	307** 315	351** 389	
Arc Resistance	LP406-4011.2	seconds	191	184***	191***	180-350***
Arc Tracking	ASTM D2303	minutes				300-600
Thermal Conductivity	ASTM C177	cal-cm/sec cm^2°C	1.40×10^{-4}	14.9×10^{-4}	14.9×10^{-4}	
Thermal Expansion Coefficient	ASTM D696	in/in/°C	1.59×10^{-5}		2.02×10^{-5}	

*ASTM D 150   **ASTM D 149   ***ASTM D 495   ****Flame retardant

Random Distribution, Cut Very Short Fibers

Property	ASTM Method	Units	Condition	Typical Value
Dielectric Constant Perpendicular to molding direction	D1531		2.5 GHz 10.0 GHz	2.38 ± .05 2.38 ± .05
Dissipation Factor	D1531		2.5 GHz 10.0 GHz	.0025 .0025
Coefficient of Thermal Expansion x 10^{-6} MD	D696	in/in/°C	-73 to 20°C 20 to 25°C 25 to 100°C	120 54 149
CMD			-73 to 20°C 20 to 25°C 25 to 100°C	49 76 54
Thermal Conductivity		cal-cm/hr cm^2°C	23 to 100°C	2.5

MD - molded direction
CMD - perpendicular to molded direction

CONCRETE PAVEMENT

Ref. 61

Sample	Density	(MHz)	0.1	1	10	100
S1	Dry	κ	9.05	7.97	7.01	6.57
		tan δ	.0946	.0913	.0730	.0536
S1	Wet	κ	176.5	69.2	23.5	13.2
		tan δ	.822	1.088	.734	.485

Concrete pavement at 40% R.H., 25°C, 14 GHz

1	0.1			Various	5.03-5.06	.026-.029
2	0.1			Various	5.06-5.17	.034-.030
3	0.335	2.14	2.21	Face 1	5.21	.059
				Face 1, 90°	5.20	.0612
				Face 2	5.30	.0509
				Face 2, 90°	5.26	.0505
4	0.453	2.04	2.81	Face 1	4.71	.0470
				Face 1, 90°	4.60	.0455
				Face 2	4.70	.0487
				Face 2, 90°	4.55	.0487

Sample	Density	(Hz)	10^5	10^6	10^7	10^8
S	Dry	κ	4.51	4.34	4.21	4.14
		tan δ	.0280	.0221	.0181	.0198
S	Wet	κ	42.0	17.7	9.03	6.54
		tan δ	.875	.638	.444	.233
L	Dry	κ	4.79	4.73	4.70	4.61
		tan δ	.0187	.0158	.0123	.0121
L	Wet	κ	14.48	9.28	6.65	6.01
		tan δ	.368	.280	.190	.104

Asphalt pavement at 40% R.H., 25°C, 14 GHz

Sample No.	Thickness (cm)	Density (g/cm^3)	H$_2$O (%)	Orientation	κ'	tan δ
1	0.1			Independent	4.73	.0114
2	0.1			Independent	4.62	.0103
3	0.1				5.03	.0120
4	0.1				5.48	.0095
5	0.91	2.35	.754	Face 1	6.02	.021
				Face 1, 90°	5.53	.052
				Face 2	5.37	.204
				Face 2, 90°	5.44	.102

REFERENCES

1. ANON. Thermoplastics for Load-Bearing Electrical Applications. MATERIALS ENGINEERING, v. 76, no. 6, June 1972. p. 46-47.

2. AIR FORCE MATERIALS LAB., MANUFACTURING TECHNOL. DIV., Wright-Patterson Air Force Base, Ohio. Manufacturing Methods for High Temperature Reinforced Plastic Aircraft Radomes. Jan. 1972.

3. APONYI, T.J. High Temperature Composite Radome Materials. U.S.A.F. - Georgia Tech. Symp. on Electromagnetic Windows. June 1966.

4. BISHAY, A. Electrical Conductivity in Glasses and Glass-Metal Composites. American Univ. in Cairo, Dept. of Materials Engineering and Physical Sci., Egypt. Nov. 24, 1971. Avail. NTIS* as AD 734 252.

5. CELANESE CORP., Morris Court, Summit, New Jersey. Advanced Engineering Composites. What's New in Materials.

6. CONTINENTAL-DIAMOND FIBRE CORP., Newark, Delaware. Polyester Glass - Dilecto and Celoron. Sept. 1965.

7. COPELAND, R.L. and V.A. CHASE. Development of Fiber Reinforced Ceramic Radomes. U.S.A.F. Georgia Tech. Symposium on Electromagnetic Windows, Volume I. June 1966.

8. DEACON, R.F. Electrical Resistivity of Boron Fibers. BOEING SCIENTIFIC PHYS. LABS., Seattle, Washington, Aug. 1967. Avail. NTIS* as AD 658 896.

9. DuPONT. DuPont Glass-Reinforced Zytel Nylon Resins. Sept. 1969.

10. DuPONT. DuPont's New High Modulus Organic Fiber for Plastics Reinforcement Ballistic Armor and Tension Cable Applications-PRD-49. Preliminary Data, Sept. 15, 1971.

11. DuPONT. Pyralin Polyimide High Temperature Resistant Materials.

12. BRELAND, J.G., JR. et al. Lightning Protective Coatings for Boron and Graphite Fiber Reinforced Plastics. In: 1970 Lightning and Static Electricity Conference, Dec. 9-11, 1970. p. 233-251.

13. BATTELLE MEMORIAL INST., Columbus, Ohio. Electrical and Thermal Transport Models for Analysis of Reinforced Composites. By: DUGA, J.J. Contract No. NONR-4925(00). July 1966. Avail. NTIS* as AD 486 667.

14. FABIAN, R.J. Engineer's Guide to Polyimide Plastics. MATERIALS ENGINEERING, v. 74, no. 2, Aug. 1971. p. 26-31.

* National Technical Information Service, Springfield, Virginia 22151

15. FARAG, M.M. et al. Some Physical and Mechanical Properties of Glass-Aluminum Metal Composites. American Univ. of Cairo, Solid State Res. Center, Egypt. Feb. 1972. Avail. NTIS* as AD 738 663.

16. FEKETE, F. et al. Electrical Properties of Novel Heat Resistant-Fast Curing Thermosetting Resin Systems. Electrical Insulation Conf., Proc., 7th, Conf., Oct. 15-19, 1967. p. 1-5.

17. FIBERFIL DIVISION, REXALL CHEMICAL CO., Evansville, Indiana. Fiberfil Reinforced Thermoplastics.

18. FIBERITE CORPORATION, Winona, Minnesota. Melamine, Phenolic, Epoxy and Polyesters. Apr. 1970.

19. MASSACHUSETTS INST. OF TECHNOL., DEPT. OF PHYSICS, Cambridge, Mass. Amorphous Carbon Films: Conduction Across Metal/Carbon/Metal Sandwiches. By: MacVICAR, M.L.A. Contract No. N00014-67-A-0204-0041. Sept. 1970. Avail. NTIS* as AD 712 073.

20. GALASSO, F.S. Unidirectionally Solidified Eutectics for Optical, Electronic, and Magnetic Applications. J. OF METALS, v. 19, no. 6, June 1967. p. 17-21.

21. GENERAL ELECTRIC COMPANY, Pittsfield, Mass. GEMON Thermoset Polyimide.

22. GENERAL ELECTRIC COMPANY, INSULATING MATERIALS DEPT., Schenectady, New York. High Temperature Polyimide Prepregs.

23. INTERAND CORP., Rockville, Md. Whisker Reinforcement of Piezoelectric Transducer Ceramics. By: FEITH, K.E. Contract No. N00014-70-C-0182. Aug. 1971. Avail. NTIS* as AD 730 484.

24. GLASTIC CORP., Cleveland, Ohio. Fiber Glass Electrical Insulation Rod. Nov. 1967.

25. GLASTIC CORP., Cleveland, Ohio. Fiber Glass Plastic Sheet Stock. Aug. 1, 1965.

26. GLASTIC CORP., Cleveland, Ohio. Fiber Glass Reinforced Plastic Structural Insulating Materials. July 1972.

27. HAWLEY PRODUCTS, A HITCO CO. Fiber Glass Reinforced Plastics.

28. HEXCEL AEROSPACE. Composite Materials. Data Sheets.

29. HOGGATT, J.T. High Performance Filament Wound Composites for Pressure Vessel Applications. National SAMPE Technical Conference, Huntsville, Alabama, Oct. 1971. p. 157-167.

30. HOLLIDAY, L. Composite Materials. ELSEVIER PUBLISHING CO., 1966.

31. NASA. Polyimide Resin-Fiberglass Cloth Laminates for Printed Circuit Boards. By: KENNEDY, B.W. Patent Application. Sept. 10, 1970. 13 p.

32. INTERAND CORP., Rockville, Md. A New Composite Ceramic Piezoelectric Transducer Material. By: LESTER, W.W. Contract No. NOO014-70-C-0182. Nov. 1970. Avail. NTIS* as AD 714 494.

33. DARMORY, F.P. et al. P13N: Polyimide Laminating Varnish. Presented at the Annual Institute of Printed Circuits Meeting, Washington, D.C., Apr. 5-8, 1971.

34. SHEWCHUN, J. and J. MITCHELL. Electrical Conduction in Silicon-Carbide Composites. IEEE PROC., v. 117, no. 10, Oct. 1970. p. 1933-1940.

35. JUN, C.K. and P.T.B. SHAFFER. Thermal Expansion of Niobium Carbide, Hafnium Carbide and Tantalum Carbide at High Temperature. J. OF THE LESS COMMON METALS, v. 24, 1971. p. 323-327.

36. HARADA, Y. and S.A. BORTZ. Properties of Hot-Pressed TaC-C and NbC-C Composites. Presented at the 69th Annual Meeting of the American Ceramic Society, May 2, 1967, New York City.

37. DAVIS, W.J. New TFE Fluorocarbon, Compound is Strong, Rigid. MATERIALS IN DESIGN ENG., v. 53, no. 3, Mar. 1961. p. 10-12.

38. HITCO, MATERIALS DIV., Gardena, Calif. Carbon and Graphite. Oct. 4, 1966.

39. HITCO, MATERIALS DIV., Gardena, Calif. Refrasil Product Bulletin, Aug. 1969.

40. HYDE, J.K. Glass Fibre Laminates in the Electrical Field. In: Glass Reinforced Plastics, Ed. by Phillip Morgan, Liffe Books, Ltd., London, 1961.

41. KNIBBS, R.H. and J.B. MORRIS. The Effects of Fibre Orientation on the Physical Properties of Composites. Plastics Inst., Conf. on Reinforced Plastics Res. Projects III, London, England, Nov. 10, 1971.

42. KNIBBS, R.H. et al. The Thermal and Electrical Properties of Carbon Fibre Uni-Direction Reinforced Epoxy Composites.

43. KUEBELER, G.C. and C.E. JORDAN. Advanced Composites....The New "Diet" Material for Structural Applications. HERCULES CHEMIST, July 1971. p. 1-10.

44. KUREHA CARBON FIBER. KUREHA CHEMICAL INDUSTRY CO., LTD. New York, N.Y. 10017.

45. LIQUID NITROGEN PROCESSING CORP., Santa Ana, Calif. Fortified Polymers.

46. LUBIN, G. Handbook of Fiberglass and Advanced Plastic Composites. Van Nostrand Reinhold Co., New York, New York, 1969.

47. McDANELS, D.L. Electrical Resistivity and Conductivity of Tungsten-Fiber-Reinforced Copper Composites. TRANS. OF THE ASM, v. 59, 1966. p. 994-997.

48. MOLZON, A.E. Electrical Properties of Plastic Materials; Data Compiled from Technical Conference Search. PLASTICS TECHNICAL EVALUATION CENTER, July 1965. Avail. NTIS* as AD 624 922.

49. MOREHOUSE, D.S. and H.A. WALTERS. Foamed Thermoplastic Microspheres in Reinforced Polyesters. SPE JOURNAL, v. 25, May 1969. p. 45-50.

50. MONSANTO CORP. Skybound 700 High Heat Resistant Polyimide Resin Technical Bulletin No. 5042C. Jan. 1970.

51. NAHILL, G.F. and R.A. QUINTUS. A New Laminate Offering Very High Insulation Resistance and Excellent Dielectric Strength. Electrical Insulation Conf., Proc., 7th, Conf., Oct. 15-19, 1967. p. 8-11.

52. NAUM, R.G. et al. Thermal Diffusivity and Thermal Conductivity of Carbon-Carbon Composites. Presented XI Therm. Conf. Albuquerque, N. Mexico, Sept. 28-Oct. 1, 1971.

53. PARR MOLDING COMPOUNDS, U.S. POLYMERIC, INC. Composite Data Sheets, dated 1966-1967.

54. PENTON, A.P. et al. Fundamental Investigations of High Intensity Electric Current Flow, Processes and Resultant Damage in Advanced Composites. In: 1970 Lightning and Static Electricity Conf., Dec. 1970. p. 253-259.

55. PETRIE, E.M. Reinforced Polymers for High-Temperature Microwave Applications. IEEE TRANS. ON ELECTRICAL INSULATION, v. EI-5, no. 1, Mar. 1970. p. 19-26.

56. ROBSON, D. et al. Some Electronic Properties of Polyacrylonitrile-Based Carbon Fibers. J. OF PHYS., D, v. 5, 1972. p. 169-179.

57. ROGERS CORP., Rogers, Conn. RT/Duroids - Reinforced Teflon. 1970.

58. SHAFFER, T.B. and C.K. JUN. The Elastic Modulus of Dense Polycrystalline Silicon Carbide. MAT. RES. BULL., 1972.

59. STRATTON, W.K. Evaluation of DuPont's High Modulus Organic Fiber PRD-49 Type 1. National SAMPE Technical Conf., Apr. 21-23, 1971. p. 325-343.

60. U.S. POLYMERIC, INC. Composite Data Sheets, dated 1966-1968.

61. WESTPHAL, W.B. and A. SILS. Dielectric Constant and Loss Data. MASS. INST. OF TECHNOL., Cambridge, Mass. Apr. 1972.

62. WOLF, S.M. Properties and Applications of Dispersion-Strengthened Metals. J. OF METALS, v. 19, no. 6, June 1967. p. 22-28.

63. DIETZ, A.G.H. Composite Engineering Laminates. THE MIT PRESS, Cambridge, Mass., 1969.

64. LOCKWOOD, P.A. Investigations of Glass Fiber-Metal Composite Materials. Contract No. NOrd-15764. Final Rept. - Nov. 1960. Avail. NTIS* as AD 274 530.

65. BRATSCHUN, W.R. et al. Uses of Ceramics in Microelectronics - A Survey. NASA SP-5097, 1971. Avail. U.S. Government Printing Office as 3300-0388.